Noise

Second Edition

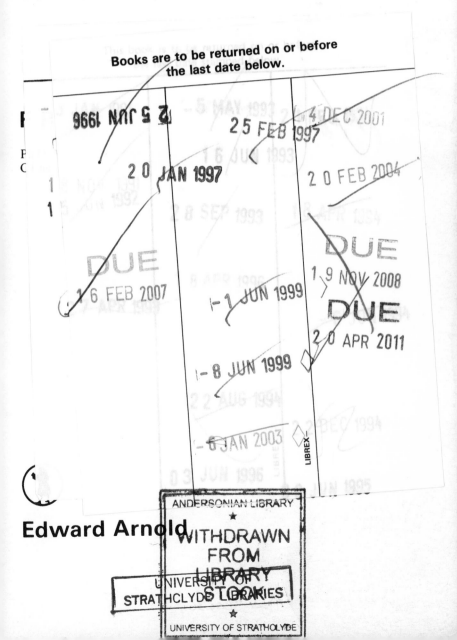

Edward Arnold

© F. R. Connor 1982

First published 1973
by Edward Arnold (Publishers) Ltd,
41 Bedford Square, London WC1B 3DQ

Edward Arnold (Australia) Pty Ltd,
80 Waverley Road, Caulfield East,
Victoria 3145, Australia

Edward Arnold
3 East Read Street, Baltimore,
Maryland 21202, U.S.A.

Reprinted 1976, 1979
Second edition 1982
Reprinted 1986

British Library Cataloguing in Publication Data

Connor, F. R.
  Noise.—2nd ed.
  1. Electronic circuits—Noise
  I. Title
  621.3815'3   TK7867.5

  ISBN 0–7131–3459–3

Photo Typeset by
Macmillan India Ltd., Bangalore

Printed in Great Britain by
Thomson Litho Ltd, East Kilbride, Scotland

M.F.F.

Preface

In this new edition, certain parts of the text have been extensively revised. A new section on random variables is introduced in Chapter 2 and some basic ideas concerning matched filtering, decision theory, and estimation theory are presented in Chapter 3. A further treatment of circuit noise is made in Chapter 4 and a new section on low-noise amplifiers is included in Chapter 5. In Chapter 6, a comparative study of the signal-to-noise performance of various systems has been extended to cover digital systems and satellite systems. As an alternative approach, the energy-to-noise density ratio and its effect on the bit error rate is also included. A further feature of the book is the extended use of appendices to cover such topics as narrowband noise, decision theory, estimation theory, and the probability of error. It is intended for the reader seeking a deeper understanding of the text and is supplemented by a large number of useful references for further reading. The book also includes several worked examples and a set of typical problems with answers.

The aim of the book is the same as in the first edition, with the difference that Higher National Certificates and Higher National Diplomas are being superseded by Higher Certificates and Higher Diplomas of the Technician Education Council.

In conclusion, the author wishes to express his gratitude to those of his readers who so kindly sent in various corrections for the earlier edition.

1982   FRC

Preface to the first edition

This is an introductory book on the important topic of *Noise*. Electrical noise is of considerable importance in communication systems and the book presents basic ideas in a coherent manner. Moreover, to assist in the understanding of these basic ideas, worked examples from past examination papers are provided to illustrate clearly the application of fundamental theory.

The book begins with a survey of the various types of electrical noise found in communication systems and this is followed by a description of some mathematical ideas concerning random variables. Circuit noise, noise factor, and noise temperature are considered in the following chapters, and the book ends with a comparative study of some important communication systems.

iv *Preface*

This book will be found useful by students preparing for London University examinations, degrees of the Council of National Academic Awards, examinations of the Council of Engineering Institutions, and for other qualifications such as Higher National Certificates, Higher National Diplomas, and certain examinations of the City and Guilds of London Institute. It will also be useful to practising engineers in industry who require a ready source of basic knowledge to help them in their applied work.

1973   FRC

Acknowledgements

The author sincerely wishes to thank the Senate of the University of London and the Council of Engineering Institutions for permission to include questions from past examination papers. The solutions and answers provided are his own and he accepts full responsibility for them.

Sincere thanks are also due to the City and Guilds of London Institute for permission to include questions from past examination papers. The Institute is in no way responsible for the solutions and answers provided.

Finally, the author wishes to thank the publishers for various useful suggestions and will be grateful to his readers for drawing his attention to any errors which may have occurred.

Abbreviations

C.E.I.   Council of Engineering Institutions examination in Communication Engineering, Part 2
C.G.L.I.   City and Guilds of London Institute examinations
U.L.   University of London, B Sc (Eng) examination in Telecommunication, Part 3

Contents

Symbols

e	electronic charge
f_i	idler frequency
f_p	pump frequency
f_s	signal frequency
g_m	forward transconductance
$h(t)$	impulse response of a filter
i_s	shot noise current
i_t	thermal noise current
k	any constant
	Boltzmann's constant
m	any number
	modulation factor (depth of modulation)
$m(t)$	modulating signal
m_f	modulation index
$n(t)$	random noise voltage
$p(x)$	continuous probability density function
q	number of quantised levels
v_t	thermal noise voltage
$\bar{x}$	average value of variable x
B	bandwidth
BER	bit error rate
C	capacitance
	communication capacity
C_i	peak input carrier power
C/N_0	carrier-to-noise power density ratio
E	energy per bit
$E[X]$	expectation or average value of variable X
$E[X^2]$	mean-squared value of variable X (variance of X)
E/N_0	signal energy-to-noise power density ratio
EIRP	effective isotropic radiated power
ENR	excess noise ratio
F	noise factor
$F(\omega)$	Fourier transform
G	power gain
G_T	power gain of transmitting antenna
G_R	power gain of receiving antenna
G/T	figure of merit
$H(\omega)$	transfer function of a network
$L[y(t)]$	likelihood ratio of $y(t)$
N_0	noise power spectral density
$P(A)$	discrete probability function of A
$P(A|B)$	conditional probability function of A given B

P_c	average carrier power
P_D	detection probability
P_F	false alarm probability
P_T	transmitted power
R	bit rate
	resistance
R_{eq}	equivalent thermal noise resistance
$R(\tau)$	autocorrelation function
$R_x(\tau)$	autocorrelation function of variable x
$R_{xy}(\tau)$	cross-correlation function of variables x and y
S	average signal power
S/N	signal-to-noise ratio
S_i/N_i	input signal-to-noise ratio
S_o/N_o	output signal-to-noise ratio
$S(\omega)$	power spectral density
T	absolute temperature
	periodic time
T_a	antenna noise temperature
T_e	effective noise temperature
T_R	receiver noise temperature
T_s	system temperature
W	energy
	highest modulating frequency
Y	Y-factor $= \dfrac{\text{noise power of standard source}}{\text{noise power of antenna}}$
α	transistor forward current gain
α_0	transistor d.c. forward current gain
$\delta(t)$	Dirac delta function
λ	wavelength
ν	frequency of events
ρ	correlation coefficient
σ	standard deviation
τ	time interval
ω	angular frequency

1

Introduction

Electrical noise can be defined as an unwanted signal which is always present in a communication system. Its presence tends to impede the reception of the wanted signal and is usually the limiting factor in its detection. Hence, the study of noise is an important part of such systems, and it is necessary to evaluate its limiting effect, since it ultimately determines the performance of the system. In addition, interference due to other unwanted signals, such as those from neighbouring stations, may also be equally important and must be considered along with noise in any general analysis.

1.1 Nature of noise

Noise and interference play a somewhat similar role in communication systems but they are dissimilar in nature in one important respect. It is usually found that noise is composed of randomly-occurring voltages which are unrelated in phase or frequency and may sometimes be of a very peaky nature. Interference, on the other hand, is usually periodic and regular in form. On close examination, noise voltages show pulse-like waveforms, some with large peaks, which occur quite randomly and continuously. When viewed on an oscilloscope, noise gives a 'spiky' impression and this is illustrated in Fig. 1.1.

On average, the peaks are about a microsecond in duration and therefore have high-frequency components. The random properties of noise require a study of its statistical behaviour, while some knowledge of its frequency and phase characteristics can be obtained by means of Fourier transform techniques.

1.2 Types of noise

As there are many sources which produce noise, they may be broadly classified as natural or artificial. Artificial or man-made noise arises mainly from electrical equipment, e.g. commutator motors, sparking plugs in ignition systems, faulty switches, electric shavers, etc. They produce 'noise-like' voltages which very often have regular properties and may be regarded also as a form of interference. The effect is impulsive in nature and usually consists of damped sine waves with a defined periodicity. The noise effect of sparking plugs is clearly seen on a domestic TV screen as a set of bright dots which cover the

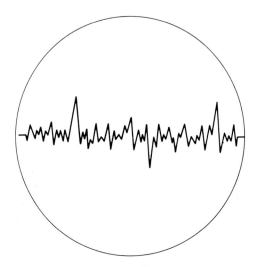

Fig. 1.1

whole picture as a car drives past the house. A study of man-made noise reveals that it can be minimised or eliminated at the source by the use of suppressors or by improved component design.

The natural forms of noise are those due to cosmic radiation, atmospherics, and the more familiar circuit noise associated with electronic circuits. Cosmic noise and atmospherics usually enter the system via the receiving antenna. The former mainly comprises electromagnetic radiation from solar and galactic sources. The study of solar radiation is a vast subject and its effect on radio reception is well known. However, its direct effect on the receiving antenna may be reduced by altering the directional properties of the antenna or by pointing it away from the sun.

The noise effect of galactic sources[1] is more widespread and arises due to emissions from stars in the galaxy. In particular, certain radio stars such as Cassiopeia emit definite and regular amounts of noise, which may be avoided by suitably pointing the antenna away from such sources. However, the overall effect of this 'sky noise' is usually expressed as a sky *noise temperature*. It is generally low compared to circuit noise, for example, and is found to vary over a wide frequency range.

From Fig. 1.2 it will be observed that the sky noise temperature is lowest over a frequency range known as the microwave band. It is referred to as a low-noise 'window' and is exploited particularly in the fields of radio astronomy and space communications.[2] Since the received signals from radio stars or space satellites are generally small, it is imperative to ensure that the background noise received

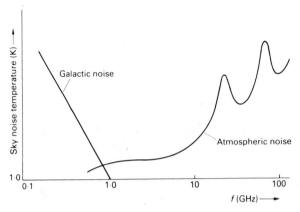

Fig. 1.2

from various parts of the sky is as low as possible. A further contribution to this background noise arises due to scattered radiation at 22 GHz, from water vapour, snow, etc., and at 60 GHz, due to O_2 absorption in the atmosphere. These effects appear as resonant peaks in Fig. 1.2.

The most troublesome form of naturally-occurring noise is circuit noise, generally known as thermal noise and shot noise. Thermal noise is produced by the random motion of 'free' electrons in a conductor. Due to collisions with vibrating atoms in the conductor, the electron motion depends on temperature and produces a fluctuating current in the conductor. The *average* current in the conductor is zero but the random fluctuations produce a noise voltage across the conductor on open-circuit. Shot noise is due to the random fluctuation in the emission of electrons from a cathode surface, which produces a shot noise current in a valve. Random variations in the diffusion of charge carriers in a semiconductor diode or transistor also give rise to a shot noise current in the device.

Circuit noise is therefore due to the nature of electronic components and cannot be eliminated. Hence, the study of circuit noise constitutes an important part of communication systems and will be dealt with in Chapter 4. However, there is some concern in finding ways to reduce it and this may be achieved by the proper choice of components and through the use of low temperatures, e.g. for both the parametric amplifier and the helium-cooled maser.

Since noise in a receiver generally covers a very wide range of frequencies, it is not possible to avoid it by choosing any particular frequency band. However, it is observed that the level of noise received within the frequency range 100 kHz to 30 MHz is fairly high and shows a decrease at higher frequencies in the VHF, UHF, and microwave bands. It tends to increase again in the millimetric region and remains somewhat high thereafter. Hence, the evaluation of the noise power in a system is based on its total value within some defined bandwidth which often is the bandwidth of the receiver used or the bandwidth of the transmission system under study.

1.3   Evaluation of noise

Mathematical concepts used in the evaluation of noise relate either to noise voltages or to noise power spectral density. Since noise fluctuations are both positive and negative, the mean-square noise voltage (or current) is first determined, which then leads to the expression of a.c. noise power $P_n = \overline{v_n^2}/R$ where $\sqrt{\overline{v_n^2}}$ is the rms noise voltage and R is the circuit resistance involved. Alternatively, it is sometimes more convenient to consider the spectral distribution of the noise present and to evaluate its spectral density $S(f)$ in a given bandwidth df. The average noise power or the mean-square noise voltage can be evaluated for a load resistance of $1\,\Omega$ by the use of integration. A particular example of this is 'white noise' which is often used as a standard of reference. Details of this are given in Chapter 3.

Since the noise power delivered to a load depends on the load resistance, it is necessary to define the exact nature of the noise power calculated. In practice, it is most usual to define this as the *available* noise power, i.e. the maximum noise power available in the load under matched conditions. This condition is most often used since it gives the worst possible effect of the noise in any given load, as can be verified by the maximum power transfer theorem, and because matched conditions generally apply in communication systems.

To obtain a better understanding of the statistical nature of noise, it is necessary to have some knowledge of probability and statistics, details of which are given in Chapter 2. A further development of these basic ideas leads to the notions of autocorrelation and cross-correlation functions, which are dealt with in Chapter 3. However, the detailed analysis of systems requires a better understanding of random signal theory[3] and decision theory.[4] The latter is particularly necessary in dealing with the extraction and detection of signals in noise, which arises mainly in the threshold areas of very weak signals, as is the case of some radar systems. In this connection, the autocorrelation and cross-correlation techniques are of greater use.

In most practical communication systems, the usual criterion of performance is the ratio of signal power to noise power in the system and is defined as the *signal-to-noise ratio* or S/N ratio at the output of the detector. This criterion is often used as a basis for comparison of different systems, as such systems may differ very widely in their particular characteristics and thus make a fair comparison difficult. As no single system can claim to be superior in every respect, very often the system chosen for a given application is a compromise between one or more alternatives.

As noise cannot be wholly eliminated from communication systems, its effect must be minimised or, alternatively, the signal-to-noise ratio (S/N) must be *maximised* as far as possible. A television system which may be regarded as good would have an S/N ratio of around 60 dB while a commercially acceptable telephone system would have an S/N ratio of around 30 dB. Nevertheless, some radar systems are still usable down to output S/N ratios as low as 3 dB, a figure which would be quite unacceptable in some other communication system such as television.

An alternative concept to the signal-to-noise ratio is that of signal energy-to-noise spectral density ratio (E/N_0) or carrier power-to-noise spectral density ratio (C/N_0). As these ratios use a bandwidth of 1 Hz only, they are independent of the actual bandwidth used in the system. Hence, they are useful for comparing the performance of systems which use widely different bandwidths. This is the approach adopted in Chapter 6 when dealing with digital and satellite communication systems.

2

Probability and statistics

In many physical phenomena, the outcome of an experiment may show random fluctuations which cannot be predicted precisely. For example, in a coin tossing experiment, it is impossible to say when heads or tails will occur. In this case, however, the outcomes over a large number of tosses may show some regularity in the results. Hence, on average we may find that heads and tails occur about evenly. The study of the *average* behaviour of events leads to a determination of the frequency of occurrence of certain outcomes such as that of heads or tails. In mathematical terms, this is defined as the notion of *probability*.[5]

Associated with probability are such concepts as probability distributions and probability density functions which are used to depict the results of a large number of events. An analysis and study of such results can show regularities which enable certain laws to be determined, and this is essentially known as *statistics*.[6]

2.1 Definition of probability

In a random experiment, such as coin tossing which is repeated several times, suppose the two outcomes are either A (heads) or B (tails). If m_A is the number of occurrences of A in a total of n tosses, the relative frequency of occurrences of A is given by m_A/n. Denoting the chance or probability of A by $P(A)$, the ratio m_A/n will approach some mean value and will show little change if n is a very large number. This ratio is defined as the probability $P(A)$ and is given by

$$P(A) = \lim_{n \to \infty} \frac{m_A}{n}$$

Similarly, if m_B is the number of occurrences of B we have

$$P(B) = \lim_{n \to \infty} \frac{m_B}{n}$$

where $P(B)$ is the probability of outcome B. Consequently, it follows that if there is a certainty of an event A occurring every time the coin is tossed then $m_A = n$ and $P(A) = 1$. Alternatively, if event A can never happen then $m_A = 0$ with $P(A) = 0$ and this signifies an impossibility. Hence, the value of any probability lies between 0 and 1, i.e. $0 \leqslant P(A) \leqslant 1$ and $0 \leqslant P(B) \leqslant 1$ with $P(A) + P(B) = 1$.

The notion of probability can be related to a point in *sample space*, since the outcome of a random experiment such as coin tossing can be related to a point in sample space and the total of all the outcomes to the whole of sample space. Such a mathematical concept enables the probability of an event to be determined in terms of sample space and leads to a study of *set theory*.[7] Further details are given in Appendix A.

2.2   Joint probability

In the case of two or more events A and B which are not mutually exclusive, i.e. they may occur together, the probability of this occurring is called the *joint* probability $P(AB)$. If m_{AB} is the number of times out of n trials that A and B occur together then we have

$$P(AB) = \lim_{n \to \infty} \frac{m_{AB}}{n}$$

In the case of events in which A or B may or may not occur together, the total probability of A or B occurring is defined as $P(A+B)$. If $(m_A + m_B)$ is the number of events of A or B out of a total of n events then

$$P(A+B) = \lim_{n \to \infty} \frac{(m_A + m_B)}{n} = \lim_{n \to \infty} \frac{m_A}{n} + \lim_{n \to \infty} \frac{m_B}{n} - \lim_{n \to \infty} \frac{m_{AB}}{n}$$

or

$$P(A+B) = P(A) + P(B) - P(AB)$$

where $P(AB)$ is the probability of A and B occurring together and must be subtracted from the sum $P(A) + P(B)$, since either $P(A)$ or $P(B)$ includes events in which A and B occur together.

2.3   Conditional probability

In certain events, the occurrence of A may depend on the prior occurrence of B. The probability of this happening is called the *conditional* probability. Hence, the conditional probability of the occurrence of A given that B has already occurred is written as $P(A|B)$ where

$$P(A|B) = m_{AB}/m_B$$

where m_{AB} is the number of joint occurrences of A and B and m_B is the number of occurrences of B (with or without A).

Hence
$$P(A|B) = \frac{m_{AB}}{n} \frac{n}{m_B} = \frac{P(AB)}{P(B)}$$

Similarly
$$P(B|A) = \frac{P(AB)}{P(A)}$$

Combining the last two equations yields

$$P(AB) = P(A|B)P(B) = P(B|A)P(A)$$

or

$$P(A|B) = \frac{P(B|A)P(A)}{P(B)}$$

which is known as Bayes' theorem.

In particular, if $P(A)$ and $P(B)$ are independent events then $P(A|B) = P(A)$ and $P(B|A) = P(B)$, hence

$$P(AB) = P(A)P(B)$$

or the *joint* probability of A and B is the product of their individual probabilities.

Comment

Two important examples of probabilities which occur in information theory[8] are known as the *a priori* probability and the *a posteriori* probability. They usually refer to the transmission and reception of signals. For example, the *a priori* probability of transmitting a signal A is simply $P(A)$ while the *a posteriori* probability of transmitting a signal A, given that some signal B has been received, is given by the *conditional* probability $P(A|B)$.

Example 2.1

Three coins are tossed at random. What is the probability (a) for all heads and (b) for all heads or all tails?

Solution
(a)   The tossing of a coin at random can result in either heads or tails and there is an equal probability for heads or tails, i.e. $P(H) = P(T) = \frac{1}{2}$. Hence, the probability of all heads in three tosses is given by the joint probability $P(HHH) = P(H)P(H)P(H) = \frac{1}{2} \times \frac{1}{2} \times \frac{1}{2} = \frac{1}{8}$.
(b)   Since the probability of all tails $P(TTT) = P(HHH) = \frac{1}{8}$, the total probability for all heads or all tails is the sum of $P(HHH)$ and $P(TTT)$, i.e. $\frac{1}{8} + \frac{1}{8} = \frac{1}{4}$.

Example 2.2

A box contains six red beads and three blue beads. Two beads are drawn out in succession. If the first bead is red, what is the probability that both beads are of different colour?

Solution
The probability of drawing out a red bead is six times out of nine or $P(R) = \frac{6}{9} = \frac{2}{3}$. Since the second bead must be blue, we require the *conditional* probability of drawing out a blue bead given that a red bead has already been drawn, i.e. $P(B|R)$. Hence, as five red beads and three blue beads remain in the box, we obtain $P(B|R) = \frac{3}{8}$.

The joint probability of drawing out the two beads is $P(RB)$ and is given by Bayes' theorem as

$$P(RB) = P(R)P(B|R) = \tfrac{2}{3} \times \tfrac{3}{8}$$

or
$$P(RB) = \tfrac{1}{4}$$

2.4 Probability functions

In the observation and analysis of discrete random processes, a random variable X is a variable quantity which may assume any discrete value and is known as a discrete random variable. For example, in a particular experiment a set of measurements of X may take on different values $x_1, x_2, x_3, \ldots, x_j$ and the probability for each is given by $P(x_1), P(x_2), \ldots, P(x_j)$. A plot of these probabilities leads to the idea of discrete probability functions shown in Fig. 2.1(a).

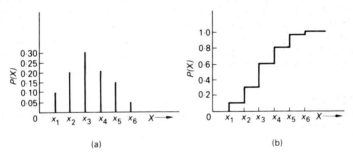

(a)   (b)

Fig. 2.1

To obtain the probability that X lies between values x_1 and x_2, it is necessary to obtain $P(x_1 < X < x_2)$ which leads to the concept of the probability *distribution* function shown in Fig. 2.1(b) where

$$P(x_1 < X < x_2) = \sum_{i=1}^{i=2} P(x_i)$$

In continuous random processes, as is observed in the fluctuations of noise voltages, the random variable X may take on a continuous set of random values and is called a continuous random variable. In this case, the concept of probability *density* function is used and is defined as $p(x)$ where

$$p(x) = \lim_{\delta x \to 0} \frac{P[x - (\delta x/2) \leqslant X \leqslant x + (\delta x/2)]}{\delta x}$$

Hence, $p(x)\,\delta x$ is the probability that X lies in the interval $[x-(\delta x/2)]$ and $[x+(\delta x/2)]$. Since the total of all probabilities of X must equal 1, we have

$$\int_{-\infty}^{+\infty} p(x)\,dx = 1$$

The probability density function is illustrated in Fig. 2.2 and its use leads to various types of distributions associated with statistical data which will be considered in Section 2.6.

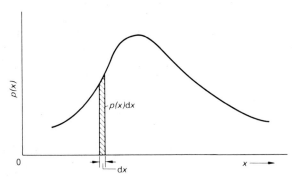

Fig. 2.2

2.5 Statistical concepts

In the analysis of statistical data, two useful concepts used are the average or expected value $\bar{x}$ of any random variable and the standard deviation σ.

Average value $\bar{x}$

For a set of discrete values $x_1, x_2, \ldots, x_j$ of the discrete random variable X with occurrences $n_1, n_2, \ldots, n_j$ respectively, the average or expected value is defined as

$$\bar{x} = E[X] = \frac{x_1 n_1 + x_2 n_2 + \ldots + x_j n_j}{n}$$

where $n = n_1 + n_2 + \ldots + n_j$.

When n is very large, n_1/n, n_2/n, etc. are the probabilities $P(x_1)$, $P(x_2), \ldots, P(x_j)$ and we obtain

$$\bar{x} = x_1 P(x_1) + x_2 P(x_2) + \ldots + x_j P(x_j)$$

or

$$\bar{x} = \sum_{i=1}^{i=j} x_i P(x_i)$$

If the random variable X is continuous, then $x_i \equiv x_j$, $P(x_i) \equiv P(x) \equiv p(x)\,\mathrm{d}x$, and the summation is from $-\infty$ to $+\infty$, hence

$$\bar{x} = E[X] = \int_{-\infty}^{+\infty} x\,p(x)\,\mathrm{d}x$$

Standard deviation σ

For a set of discrete values $x_1, x_2, \ldots, x_j$, whose average value is $\bar{x}$, their differences from the average value are $(x_1 - \bar{x})$, $(x_2 - \bar{x})$, etc. These differences are called the *deviations* and when squared are always positive. Since, in mechanics, particles of unit mass can be placed at these distances, the squares of the deviations correspond to the second moment of each unit mass about the centre of mass $(\bar{x})$. Hence, if M_2 is the second central moment we have

$$M_2 = E[(X - \bar{x})^2] = \frac{(x_1 - \bar{x})^2 + (x_2 - \bar{x})^2 + \ldots + (x_n - \bar{x})^2}{n}$$

or $$M_2 = E[(X - \bar{x})^2] = \int_{-\infty}^{+\infty} (x - \bar{x})^2 p(x)\,\mathrm{d}x$$

for the case of a continuous random variable X.

The quantity M_2 is also called the variance σ^2 while the standard deviation is defined as σ and is the square root of the variance. The standard deviation is important in statistics because it gives the *spread* of values about the average, since the smaller the value of σ, the closer are the values to the average. In a.c. theory, σ corresponds to the rms voltage or current in the circuit, as the average or expected value $\bar{x}$ is usually zero.

Example 2.3

Two random variables x and y have standard deviations σ_x and σ_y respectively. Determine the variance of their sum.

Solution

If $z = x + y$, the variance σ_z^2 is given by

$$\sigma_z^2 = \overline{(z - \bar{z})^2} = \overline{\left[(x + y) - \overline{(x + y)}\right]^2}$$

$$= \overline{\left[(x - \bar{x}) + (y - \bar{y})\right]^2} \quad \text{since} \quad \overline{(x + y)} = \bar{x} + \bar{y}$$

$$= \overline{(x - \bar{x})^2} + 2\overline{(x - \bar{x})(y - \bar{y})} + \overline{(y - \bar{y})^2}$$

Now $$\sigma_x^2 = \overline{(x - \bar{x})^2}$$

$$\sigma_y^2 = \overline{(y - \bar{y})^2}$$

and $$\overline{(x - \bar{x})} = \bar{x} - \bar{x} = 0$$

$$\overline{(y - \bar{y})} = \bar{y} - \bar{y} = 0$$

Hence $$\sigma_z^2 = \sigma_x^2 + \sigma_y^2$$

2.6   Probability distributions

In analysing a large number of statistical results, some kind of order can be obtained from the apparent randomness of the values. The values usually conform to some regular pattern called a distribution function. The most important distribution functions are the binomial distribution, Poisson distribution, and normal or Gaussian distribution.

Binomial distribution
Its main application for communication engineers lies in estimating digital errors in messages such as those transmitted by PCM repeaters. It can also be used to determine the reliability of components and it is associated with the binomial theorem.

Consider the case of a random variable X which can have two states A and B such as 0 and 1. Let the probability of A occurring be $P(A) = p$ and that of B occurring be $P(B) = (1 - p) = q$. If the experiment is repeated n times, the probability of A occurring m times is given by $P_m(A)$ where[9]

$$P_m(A) = \frac{n!}{m!(n-m)!} \, p^m q^{n-m}$$

Since the expression for $P_m(A)$ represents the mth term in the binomial expansion of $(p+q)^n$, this discrete probability function when plotted for the various possible values of m, i.e. $m = 0, 1, 2, \ldots, n$, is called the binomial distribution. A typical plot is shown in Fig. 2.3 where it is observed that the maximum probability occurs for $m = np$. This corresponds to the average value $\bar{x}$ while the standard deviation $\sigma = \sqrt{npq}$. The distribution can be applied to samples of a definite size since the number n is known.

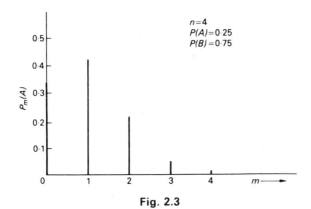

Fig. 2.3

Poisson distribution

This applies to various random phenomena, such as shot noise in a diode, the transmission of telegraph signals, and radioactive decay. It is a discrete probability distribution which can be obtained as an approximation to the binomial distribution when n is very large and p is small.

Alternatively, it can be used to evaluate the probability of an event occurring in a time interval T when it is known that v events occur per second. In this case, let the time interval T be subdivided into a large number n of smaller intervals δT where $n = T/\delta T$.

If p is the probability of an event occurring in the small interval δT, the expected number of occurrences in the time interval T is $pn = pT/\delta T$ and the number occurring per second is $pn/T = v$. Hence, $pn = pT/\delta T = vT$ and as $\delta T \to 0$, $n \to \infty$ and $p \to 0$ so that $pn = m$ is still finite. This means that in the approximate binomial distribution vT replaces pn as $\delta T \to 0$.

The probability of an event A occurring m times is given by the binomial distribution as

$$P_m(A) = \frac{n!}{m!\,n - m!}\, p^m q^{n-m}$$

$$= \frac{n(n-1)\ldots(n-m!)}{m!\,n-m!} p^m (1-p)^{n-m}$$

$$\simeq \frac{n^m}{m!}\, p^m (1-p)^n \qquad \text{(if } n \gg m)$$

$$\simeq \frac{(pn)^m}{m!}\left(1 - pn + \frac{n(n-1)}{2!}\, p^2 - \ldots\right)$$

If $p \ll 1$, the expansion in brackets is $\simeq e^{-pn}$, hence

$$P_m(A) \simeq \frac{(pn)^m}{m!}\, e^{-pn}$$

or

$$P_m(A) \simeq \frac{(vT)^m}{m!}\, e^{-vT}$$

if

$$\frac{pn}{T} = v$$

which gives the Poisson distribution with $pn = vT$. A plot of $vT = 3$ is shown in Fig. 2.4.

Normal or Gaussian distribution

The most naturally-occurring distribution is the normal or Gaussian distribution which applies to many physical phenomena, e.g. white noise, errors in practical measurements, quality control of components, etc.

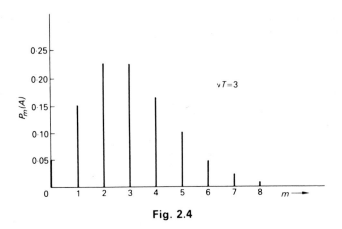

Fig. 2.4

For a continuous random variable X, the normal probability density function $p(x)$ is given by

$$p(x) = \frac{1}{\sigma\sqrt{2\pi}}\, e^{-(x-m)^2/2\sigma^2}$$

which has a most probable value at $x = m$ where m is the mean value and σ is the standard deviation. The curve is of the well-known bell shape with symmetry about $x = m$. This implies that there is an equal probability of a random value falling below or above $x = m$. It is illustrated in Fig. 2.5(a).

The parameters are such that the total area under the curve is 1, i.e.

$$\int_{-\infty}^{+\infty} p(x)\,\mathrm{d}x = 1$$

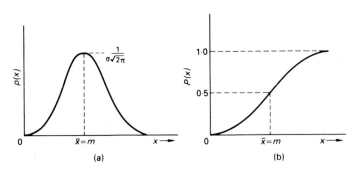

Fig. 2.5

which gives a peak value at the centre of $p(m) = 1/\sigma\sqrt{2\pi}$. The first moment of x is in fact the average or mean value since we have

$$\bar{x} = m = \int_{-\infty}^{+\infty} x p(x)\,dx = \frac{1}{\sigma\sqrt{2\pi}} \int_{-\infty}^{+\infty} x e^{-(x-m)^2/2\sigma^2}\,dx$$

To integrate this expression, introduce a dummy variable u and let $u = (x-m)/\sqrt{2\sigma^2}$ with $dx = du\sqrt{2\sigma^2}$. Hence

$$\bar{x} = \frac{1}{\sqrt{\pi}} \int_{-\infty}^{+\infty} (u\sqrt{2\sigma^2} + m)e^{-u^2}\,du$$

$$= \sigma\sqrt{\frac{2}{\pi}} \int_{-\infty}^{+\infty} u e^{-u^2}\,du + \frac{2m}{\sqrt{\pi}} \int_{0}^{\infty} e^{-u^2}\,du$$

The first integral is zero since it represents equal positive and negative areas, while the second integral is tabulated and equals $\sqrt{\pi}/2$. Hence

$$\bar{x} = \frac{2m}{\sqrt{\pi}} \times \frac{\sqrt{\pi}}{2} = m$$

The probability that the random variable X lies below some value $x = x_1$ is given by the probability distribution function shown in Fig. 2.5(b) where

$$P(x \leqslant x_1) = \int_{0}^{x_1} \frac{1}{\sigma\sqrt{2\pi}} e^{-(x-m)^2/2\sigma^2}\,dx$$

If $x_1 = m$, the mean value, we obtain

$$P(x \leqslant m) = 1/2$$

since it represents half the area under the bell-shaped curve. Hence, in general, if $-x_1 < x < +x_1$ we have

$$P(-x_1 < x < +x_1) = \frac{1}{\sqrt{\pi}} \int_{-x_1}^{+x_1} e^{-u^2}\,du = \frac{2}{\sqrt{\pi}} \int_{0}^{x_1} e^{-u^2}\,du$$

and the integral is called the *error function* of x or, abbreviated, erf x. Hence

$$\text{erf } x = \frac{2}{\sqrt{\pi}} \int_{0}^{x} e^{-u^2}\,du$$

which is tabulated in Appendix B for various values of x.

Rayleigh's distribution
This probability distribution occurs in the study of band-limited Gaussian noise and in short-term fading due to tropospheric scattering.[10] In the ballistics

field, it gives the probability of hitting a target area because it is associated with a probability distribution in two dimensions.

Suppose x and y are two independent variables, each with a normal distribution, variance σ^2, and zero mean. If (x, y) represents a point in the xy plane then for polar coordinates (r, θ) we have $r^2 = x^2 + y^2$ and $\tan\theta = y/x$. Hence, the joint probability function $p(x, y)\,dx\,dy$ transforms to $p(r, \theta)\,dr\,d\theta$ and we obtain

$$p(x, y)\,dx\,dy = p(r, \theta)\,dr\,d\theta$$

Now
$$p(x, y) = p(x)p(y) = \frac{1}{\sigma\sqrt{2\pi}}\,e^{-x^2/2\sigma^2}\,\frac{1}{\sigma\sqrt{2\pi}}\,e^{-y^2/2\sigma^2}$$

or
$$p(x, y) = \frac{1}{2\pi\sigma^2}\,e^{-(x^2+y^2)/2\sigma^2}$$

and transforming to polar coordinates then yields

$$p(r, \theta) = \frac{r}{2\pi\sigma^2}\,e^{-r^2/2\sigma^2}$$

with
$$p(r, \theta) = p(r)p(\theta)$$

since r and θ are independent variables with $0 < r < \infty$ and $0 < \theta < 2\pi$. The total probability in the angular direction is 1 and so, from circular symmetry, the probability in any direction θ is $p(\theta) = 1/2\pi$ since θ varies from 0 to 2π. Hence

$$p(r) = \frac{p(r, \theta)}{p(\theta)} = \frac{r}{2\pi\sigma^2}\,e^{-r^2/2\sigma^2}/(1/2\pi)$$

or
$$p(r) = \frac{r}{\sigma^2}\,e^{-r^2/2\sigma^2}$$

which is **Rayleigh's distribution** and is shown in Fig. 2.6.

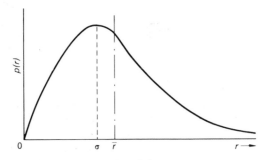

Fig. 2.6

The mean value can be shown to be $\bar{r} = \sigma\sqrt{\pi/2}$ and the standard deviation is $\sigma = 0\cdot798\bar{r}$. The probability distribution function $P(r \leqslant r_1)$ is given by

$$P(r \leqslant r_1) = \int_0^{r_1} p(r)\,dr = 1 - e^{-r_1^2/2\sigma^2}$$

Example 2.4
Band-limited white noise which has a Rayleigh distribution is applied to the input of a quadratic detector whose output voltage $v = ku^2$ where u is the input voltage and k is a constant of proportionality. Determine the probability distribution of v.

Solution
Since the input voltage has a Rayleigh distribution it is given by

$$p(u) = \frac{u}{\sigma^2} e^{-u^2/2\sigma^2} \qquad (u \geqslant 0)$$

where σ^2 is the variance. Moreover, as probabilities transform directly we must have

$$p(v)\,dv = p(u)\,du$$

or

$$p(v) = p(u)(du/dv)$$

Since

$$v = ku^2$$

$$dv/du = 2ku$$

or

$$du/dv = 1/2ku$$

Hence

$$p(v) = \frac{u}{\sigma^2} e^{-u^2/2\sigma^2} \frac{1}{2ku} \qquad (u \geqslant 0)$$

or

$$p(v) = \frac{e^{-u^2/2\sigma^2}}{2k\sigma^2}$$

2.7 Random processes[3,9]

In randomly-varying phenomena, e.g. electrical noise, earth tremors, or wind gusts, the variation with time can be represented by sets of waveforms as shown in Fig. 2.7. Each waveform at any position of time t represents a random variable X whose value varies randomly from instant to instant because it is unpredictable. Every such waveform represents a *sample* function of the random or stochastic process and the set of waveforms, $X_1(t)$, $X_2(t)$, etc., is known as an *ensemble*.

To ascertain any statistical relationship in the random process, various mathematical operations can be performed over the sample functions using ensemble averaging statistics or time averaging statistics. In ensemble averaging, the various values, $X_1(t_0)$, $X_2(t_0)$, $X_3(t_0)$, etc., of the random variable X at a relative position in time t_0 are averaged *across* the ensemble of waveforms

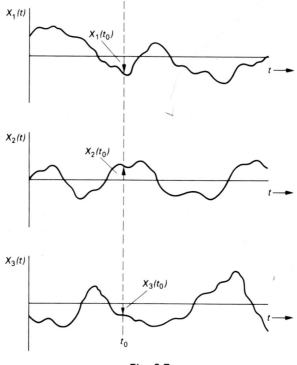

Fig. 2.7

whereas, in time averaging, a particular sample function only is considered, e.g. $X_1(t)$, and averaging is performed over a time interval of this sample. Two of the most useful quantities are the average or expected value $E[X]$ and the autocorrelation function $R_x(\tau)$ (which is discussed in the next chapter).

Stationarity

A random process in which the average value of the random variable X is the same using ensemble averaging at any time $t = \tau_1$ or $t = \tau_2$ is called a *stationary* process as its statistics are independent of the particular time position chosen. For such a process, the average or expected value is given by

$$E[X] = \int_{-\infty}^{+\infty} x\, p(x)\, \mathrm{d}x$$

where $p(x)$ is the probability density function of the random variable X in the interval $0 < x < x + \mathrm{d}x$. Similarly, the autocorrelation function $R_x(\tau)$ over the time interval $\tau = (\tau_2 - \tau_1)$ is given by

$$E[X_1(\tau_1)X_1(\tau_2)] = R_x(\tau) = \int_{-\infty}^{+\infty} \int_{-\infty}^{+\infty} x_1 x_2 p(x_1, x_2)\, dx_1\, dx_2$$

where $p(x_1, x_2)$ is the joint probability density function of the random variable X.

If the average value and the autocorrelation function are both independent of time, the process is stationary in the *strict* sense. However, if the autocorrelation function depends on the time interval $\tau = (\tau_2 - \tau_1)$, then it is stationary only in the *wide* sense.

Ergodicity

A random process in which ensemble averaging leads to the same result as time averaging is a stationary process which is also called an *ergodic* process. Hence, stationarity is an essential property of an ergodic process, but not all stationary processes are ergodic. Moreover, in an ergodic process, its statistical behaviour can be ascertained by a *single* sample function and hence the average value $E[X]$, mean-square value $E[X^2]$, and autocorrelation function $R_x(\tau)$ are given by

$$E[X] = \lim_{T \to \infty} \frac{1}{T} \int_{-T/2}^{+T/2} x(t)\, dt$$

$$E[X^2] = \lim_{T \to \infty} \frac{1}{T} \int_{-T/2}^{+T/2} x^2(t)\, dt$$

$$R_x(\tau) = \lim_{T \to \infty} \frac{1}{T} \int_{-T/2}^{+T/2} x(t)x(t - \tau)\, dt$$

Typical examples

Random processes occurring in electrical systems are thermal noise in a resistor or shot noise in a transistor, random binary transmission of digital signals, or the random telegraph signal. A particular example of electrical noise is called *white noise*, further details of which are given in the next two chapters.

Power spectral density

If the random process is ergodic, a further relationship between the power spectral density function $S(\omega)$ and the autocorrelation function $R_x(\tau)$ can be obtained. The power spectral density function and autocorrelation function form a Fourier transform pair which is given by the Wiener–Khintchine theorem. Further details of this theorem are given in Section 3.5.

3

Correlation techniques

In the study of many random processes, e.g. noise, earthquakes, etc., it is of interest to know whether there is any statistical regularity or *correlation* in the random process. The possibility of obtaining such a correlation leads to the use of various correlation techniques[11] for studying random signals such as noise.

3.1 Correlation coefficient

In the analysis of experimental data involving two random variables x and y, a plot of these two variables as shown in Fig. 3.1 may indicate some interdependence or correlation between them. For example, x may represent the age of different persons and y their respective incomes and it may be observed, generally, that the income y of people increases with increasing age x. A well-known concept used for defining any such correlation is the *correlation coefficient* ρ.

It is obvious from Fig. 3.1 that there is some linear relationship between x and y. A measure of the closeness of the values of x and y from the straight line $y = mx + c$ is obtained by evaluating the correlation coefficient ρ. Its normal-

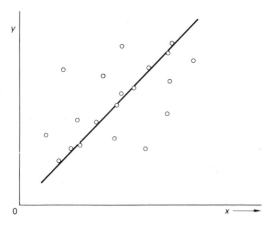

Fig. 3.1

ised value when x and y have zero mean values is given by

$$\rho = \frac{\overline{xy}}{[\overline{x^2}\,\overline{y^2}]^{\frac{1}{2}}} = \frac{\overline{xy}}{\sigma_x \sigma_y}$$

where $\overline{xy}$ is the mean of the product of the variables and σ_x and σ_y are the standard deviations of x and y respectively. The value of ρ is such that $-1 \leqslant \rho \leqslant +1$ and it is independent of the scale factors of x and y.

If $\rho = +1$ or -1, there is complete correlation and the points all fall on the straight line $y = mx + c$. If $\rho = 0$, there is no correlation, i.e. the points are scattered all around the straight line, and the variables are then considered as *orthogonal*.

Example 3.1
Two variates x_1 and x_2 have a correlation coefficient ρ and variances of σ_1^2 and σ_2^2 respectively. Hence, determine the variance of $(x_1 + kx_2)$, if k is a constant.

Solution
For $y = (x_1 + kx_2)$ with variance σ_y^2 we have

$$\sigma_y^2 = \overline{(y - \bar{y})^2} = \overline{[(x_1 + kx_2) - \overline{(x_1 + kx_2)}]}^2$$

$$= \overline{[(x_1 - \bar{x}_1) + k(x_2 - \bar{x}_2)]}^2 \quad \text{since} \quad \overline{x_1 + kx_2} = \bar{x}_1 + k\bar{x}_2$$

$$= \overline{[(x_1 - \bar{x}) + k(x_2 - \bar{x})]}^2 \quad \text{since} \quad \bar{x}_1 = \bar{x}_2 = \bar{x}$$

or $\quad \sigma_y^2 = \overline{(x_1 - \bar{x})^2} + 2k\overline{(x_1 - \bar{x})(x_2 - \bar{x})} + k^2\overline{(x_2 - \bar{x})^2}$

$$= \sigma_1^2 + 2k\overline{(x_1 - \bar{x})(x_2 - \bar{x})} + k^2\sigma_2^2$$

Now $\quad \rho = \dfrac{\overline{(x_1 - \bar{x})(x_2 - \bar{x})}}{\sigma_1 \sigma_2} \quad$ (assuming $\bar{x} \neq 0$)

Hence $\quad \sigma_y^2 = \sigma_1^2 + 2k\rho\sigma_1\sigma_2 + k^2\sigma_2^2$

Example 3.2
Two signals are defined by $v_1 = V_1 \sin \omega_c t$ and $v_2 = V_2 \cos \omega_c t$. Show that there is no correlation between them.

Solution

$$v_1 = V_1 \sin \omega_c t \quad \text{with} \quad \sigma_1 = V_1/\sqrt{2}$$

$$v_2 = V_2 \cos \omega_c t \quad \text{with} \quad \sigma_2 = V_2/\sqrt{2}$$

Hence $\quad \overline{v_1 v_2} = \overline{V_1 V_2 \sin \omega_c t \cos \omega_c t} = \dfrac{V_1 V_2}{2} \overline{\sin 2\omega_c t} = 0$

or
$$\overline{v_1 v_2} = 0$$

Now
$$\rho = \frac{\overline{v_1 v_2}}{\sigma_1 \sigma_2} = \frac{0}{\sigma_1 \sigma_2}$$

Hence
$$\rho = 0$$

3.2   Autocorrelation function

The possibility of correlation between two random variables such as $x(t)$ and $X(t)$ leads to the notion of the *autocorrelation* function which endeavours to correlate a random signal $x(t)$ with *itself* but delayed by a time interval τ, i.e. $x(t - \tau)$. The autocorrelation function $R_x(\tau)$ of the signal $x(t)$ is defined as

$$R_x(\tau) = \lim_{T \to \infty} \frac{1}{T} \int_{-T/2}^{+T/2} x(t)x(t - \tau)\,dt$$

where the definition applies to all forms of signals either random or periodic (deterministic).

The process of finding $R_x(\tau)$ implies the multiplication of $x(t)$ with the delayed signal $x(t - \tau)$ and averaging the result over the time interval T. Since τ is variable, it leads to different values of $R_x(\tau)$. Of particular interest is the value $R_x(0)$ when $\tau = 0$. In this case, we obtain

$$R_x(0) = \lim_{T \to \infty} \frac{1}{T} \int_{-T/2}^{+T/2} x(t)x(t)\,dt$$

or
$$R_x(0) = \lim_{T \to \infty} \frac{1}{T} \int_{-T/2}^{+T/2} x^2(t)\,dt$$

which will be shown later to be related to the power spectral density function $S(f)$.

Comments
1. $R_x(\tau) = R_x(-\tau)$ and the function is an even function.
2. $R_x(\tau) < R_x(0)$ and so $R_x(0)$ is the maximum value of the function $R_x(\tau)$.

An example of the graphical meaning of the autocorrelation function is shown in Fig. 3.2 for the case of an asymmetrical square wave delayed by an amount τ.

In this example, as $x(t)$ is a periodic signal, $R_x(\tau)$ is essentially the mean height of $x(t)x(t - \tau)$ taken over a period T as the waveform repeats itself after a period T. By overlapping the two waveforms $x(t)$ and $x(t - \tau)$, as shown in Fig. 3.2, we can determine the value of $x(t)x(t - \tau)$ by inspection. Hence, when $0 < \tau \leqslant T/2$

$$R_x(\tau) = \lim_{T \to \infty} \frac{1}{T} \int_{-T/2}^{+T/2} x(t)x(t - \tau)\,dt = \frac{1}{T} \int_{-T/2}^{+T/2} x(t)x(t - \tau)\,dt$$

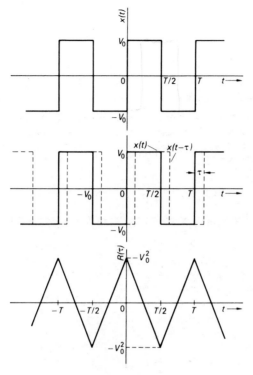

Fig. 3.2

since the function $x(t)$ is periodic. Hence

$$R_x(\tau) = \frac{1}{T}\left[\int_{-T/2}^{-T/2+\tau}(-V_0)\,V_0\,\mathrm{d}t + \int_{-T/2+\tau}^{0}(-V_0)(-V_0)\,\mathrm{d}t\right.$$

$$\left. + \int_0^{\tau}V_0(-V_0)\,\mathrm{d}t + \int_{\tau}^{+T/2}V_0\,(V_0)\,\mathrm{d}t\right]$$

$$= \frac{1}{T}\left[-V_0^2\left(-\frac{T}{2}+\tau+\frac{T}{2}\right) + V_0^2\left(\frac{T}{2}-\tau\right)\right.$$

$$\left. - V_0^2(\tau) + V_0^2\left(\frac{T}{2}-\tau\right)\right]$$

$$= \frac{V_0^2}{T}(T-4\tau)$$

or $$R_x(\tau) = V_0^2(1 - 4\tau/T) \quad \text{with} \quad R(0) = V_0^2$$

Since $R_x(\tau) = R_x(-\tau)$, the waveform of $R_x(\tau)$ between $-T/2 \leqslant \tau < 0$ is the same as that between $0 < \tau \leqslant T/2$ and is the sawtooth waveform shown in Fig. 3.2 which repeats itself after a period T as $x(t)$ is a periodic function.

Example 3.3

Show that the autocorrelation function of the periodic signal $v(t) = V_0 \sin(\omega t - \phi)$ where $0 < \phi < 2\pi$ is also periodic and determine its maximum value.

Solution

The autocorrelation function $R(\tau)$ is given by

$$R(\tau) = \lim_{T \to \infty} \frac{1}{T} \int_{-T/2}^{+T/2} x(t)x(t-\tau)\,dt$$

$$= \frac{1}{T} \int_{-T/2}^{+T/2} V_0 \sin(\omega t - \phi)V_0 \sin\{\omega(t-\tau) - \phi\}\,dt$$

$$= \frac{V_0^2}{2T} \int_{-T/2}^{+T/2} 2 \sin(\omega t - \phi)\sin\{\omega(t-\tau) - \phi\}\,dt$$

or $$R(\tau) = \frac{V_0^2}{2T} \int_{-T/2}^{+T/2} [\cos\omega\tau - \cos\{\omega(2t-\tau) - 2\phi\}]\,dt$$

$$= \frac{V_0^2}{2T} \int_{-T/2}^{+T/2} \cos\omega\tau\,dt$$

since the integral of the second cosine function over a period T is zero. Hence

$$R(\tau) = \frac{V_0^2}{2T} \cos\omega\tau \times T = \frac{V_0^2}{2} \cos\omega\tau$$

The maximum value of $R(\tau)$ is $R(0) = V_0^2/2$ and so $R(\tau)$ is also periodic with $\tau = 2\pi/\omega$. It is shown in Fig. 3.3.

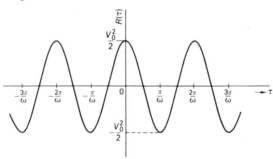

Fig. 3.3

Example 3.4
Derive an expression for the autocorrelation function of the random telegraph signal shown in Fig. 3.4, assuming that the time instants of the zero crossings follow a Poisson distribution.

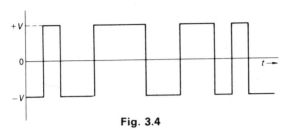

Fig. 3.4

Solution
Assume that the signal changes between values of $\pm V$ through the zeros. For m changes in the time interval T, the probability of crossing a zero is $P_m(0)$ which is more conveniently written as $P(m)$. Hence, we have

$$P(m) = \frac{(vT)^m}{m!}e^{-vT}$$

where $v = m/T$ is the *average* number of 'zero' changes per second.

To evaluate the autocorrelation function we require to know that $x(t)x(t-\tau) = \pm V^2$ where the plus sign is used for an even number of zero changes and the minus sign for an odd number of zero changes in the interval T. Hence

$$R(\tau) = \lim_{T \to \infty} \frac{1}{T} \int_{-T/2}^{+T/2} x(t)x(t-\tau)\,dt = \text{average value of } [x(t)x(t-\tau)]$$

Now the average value of any variable x was given in Section 2.5 as

$$\bar{x} = \sum_{i=1}^{i=j} x_i P(x_i)$$

Here the variable is the product $x(t)x(t-\tau)$ with values $\pm V^2$ depending on the number of changes m in the time interval T. Hence, we obtain

$$R(\tau) = \sum_{m \text{ even}} V^2 P(m) - \sum_{m \text{ odd}} V^2 P(m) = V^2 \left[\sum_{m \text{ even}} P(m) - \sum_{m \text{ odd}} P(m) \right]$$

Since the probabilities $P(m)$ are true for any interval of time T, setting $T = \tau$ yields

$$R(\tau) = V^2[P(0) + P(2) + \ldots] - V^2[P(1) + P(3) + \ldots]$$

$$= V^2 e^{-v\tau}\left[1 + \frac{(v\tau)^2}{2!} + \ldots\right] - V^2 e^{-v\tau}\left[v\tau + \frac{(v\tau)^3}{3!} + \ldots\right]$$

$$= V^2 e^{-v\tau}\left[1 - v\tau + \frac{(v\tau)^2}{2!} - \ldots\right]$$

$$= V^2 e^{-v\tau}[e^{-v\tau}]$$

or $\qquad R(\tau) = V^2 e^{-2v\tau}$

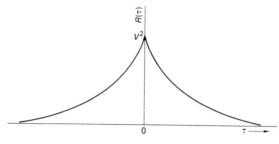

Fig. 3.5

and is shown in Fig. 3.5. Since

$$R(\tau) = R(-\tau)$$

more generally

$$R(\tau) = V^2 e^{-2v|\tau|}$$

Comment
We observe that the curve in Fig. 3.5 is narrow at low values of τ; the narrower is $R(\tau)$, the greater is the number of changes in the interval $T = \tau$, indicating the presence of high-frequency components. This will be confirmed later when the corresponding spectral function $S(f)$ is obtained for this signal.

3.3   Cross-correlation function

Analogous to the autocorrelation function, the possible correlation or statis-tical dependence between two *different* random variables $x(t)$ and $y(t)$ is expressed by the cross-correlation function $R_{xy}(\tau)$ where

$$R_{xy}(\tau) = \lim_{T \to \infty} \frac{1}{T} \int_{-T/2}^{+T/2} x(t)y(t-\tau)\,\mathrm{d}t$$

This function finds application in communication systems[11,12] and in control systems. In the former case, it reveals the transmission properties of a channel and leads to a knowledge of intersymbol interference or crosstalk.

Comments
1. $R_{xy}(\tau) = R_{yx}(\tau)$ though, generally, $R_{xy}(\tau) \neq R_{xy}(-\tau)$.
2. $R_{xy}(\tau) \leqslant \sqrt{R_x(0)R_y(0)}$.
3. $R_{xy}(\tau) = 0$ signifies that the random processes $x(t)$ and $y(t)$ are statistically independent.

Example 3.5
Two random signals are given by $m(t) = A \sin(\omega_1 t + \theta)$ and $n(t) = B \sin(\omega_2 t + \phi)$ where A and B are constants and θ and ϕ vary uniformly from 0 to 2π. Show that the cross-correlation function $R_{mn}(\tau) = 0$.

Solution

$$R_{mn}(\tau) = \lim_{T \to \infty} \frac{1}{T} \int_{-T/2}^{+T/2} m(t)n(t-\tau)\, dt$$

$$= \frac{1}{T} \int_{-T/2}^{+T/2} A \sin(\omega_1 t + \theta) B \sin\{\omega_2 t - (\omega_2 \tau - \phi)\}\, dt$$

$$= \frac{AB}{T} \int_{-T/2}^{+T/2} [(\sin \omega_1 t \cos \theta + \cos \omega_1 t \sin \theta)$$

$$\times \{\sin \omega_2 t \cos(\omega_2 \tau - \phi) - \cos \omega_2 t \sin(\omega_2 \tau - \phi)\}]\, dt$$

or

$$R_{mn}(\tau) = AB\left[\cos \theta \cos(\omega_2 \tau - \phi)\frac{1}{T} \int_{-T/2}^{+T/2} \sin \omega_1 t \sin \omega_2 t\, dt \right.$$

$$+ \sin \theta \cos(\omega_2 \tau - \phi)\frac{1}{T} \int_{-T/2}^{+T/2} \cos \omega_1 t \sin \omega_2 t\, dt$$

$$- \cos \theta \sin(\omega_2 \tau - \phi)\frac{1}{T} \int_{-T/2}^{+T/2} \sin \omega_1 t \cos \omega_2 t\, dt$$

$$\left. - \sin \theta \sin(\omega_2 \tau - \phi)\frac{1}{T} \int_{-T/2}^{+T/2} \cos \omega_1 t \cos \omega_2 t\, dt \right]$$

Each of the integrals is of standard form and is equal to zero when $\omega_1 \neq \omega_2$, hence

$$R_{mn}(\tau) = 0$$

3.4 Power spectral density

To obtain the average power associated with a periodic or random signal $x(t)$, which extends over an indefinite period of time t, it is assumed that the average power must be finite. The evaluation is based on the *power* spectral density function $S(f)$ which is *defined* as the average power at frequency f in a $1\,\Omega$ load resistor. Hence, $S(f)\, df$ gives the average power in the frequency range df and so the average power P_{av} in the periodic or random signal $x(t)$ is given by

$$P_{av} = \int_{-\infty}^{+\infty} S(f)\, df = \frac{1}{2\pi} \int_{-\infty}^{+\infty} S(\omega)\, d\omega$$

Comments
1. When $x(t)$ is a *periodic* signal, it is shown in Appendix C that $S(\omega)$ is associated with the Fourier transform $F(\omega)$ of $x(t)$ and is given by

$$S(\omega) = \lim_{T \to \infty} \frac{|F(\omega)|^2}{T}$$

2. Since $|F(\omega)|^2$ represents energy, it increases as $T \to \infty$ but, in the limit, the ratio $|F(\omega)|^2/T$ is assumed to be finite and so $S(\omega)$ can be defined as a finite quantity.
3. When $x(t)$ is a *random* signal, the expression for $S(\omega)$ becomes

$$S(\omega) = \lim_{T \to \infty} \frac{\overline{|F(\omega)|^2}}{T}$$

where $\overline{|F(\omega)|^2}$ is the ensemble average. However, if the random process is ergodic then $|F(\omega)|^2$ is the same as $\overline{|F(\omega)|^2}$.

3.5   Wiener–Khintchine theorem[13,14]

An important relationship between the autocorrelation function $R(\tau)$ and the spectral density function $S(f)$ is given by this theorem which states that $R(\tau)$ and $S(f)$ form a set of Fourier transform pairs given by

$$S(f) = \int_{-\infty}^{+\infty} R(\tau)e^{-j\omega\tau}\,d\tau$$

$$R(\tau) = \int_{-\infty}^{+\infty} S(f)e^{j\omega\tau}\,df$$

and the proof of this theorem is given in Appendix D.

Example 3.6
With the aid of the Wiener–Khintchine theorem, determine the spectral density function $S(f)$ of the random telegraph signal of Example 3.4. What is the physical significance of the result?

Solution
The autocorrelation function of the random telegraph signal was previously obtained as

$$R(\tau) = V^2 e^{-2v|\tau|}$$

Also, we have

$$S(f) = \int_{-\infty}^{+\infty} R(\tau)e^{-j\omega\tau}\,d\tau$$

$$= \int_{-\infty}^{+\infty} V^2 e^{-2v|\tau|}\, e^{-j\omega\tau}\,d\tau$$

or

$$S(f) = V^2 \left[\int_{-\infty}^{0} e^{-2v|\tau|}\, e^{-j\omega\tau}\,d\tau + \int_{0}^{\infty} e^{-2v|\tau|}\, e^{-j\omega\tau}\,d\tau \right]$$

$$= V^2 \left[\int_{0}^{\infty} e^{-2v|\tau|}\, e^{j\omega\tau}\,d\tau + \int_{0}^{\infty} e^{-2v|\tau|}\, e^{-j\omega\tau}\,d\tau \right]$$

$$= V^2 \int_0^\infty e^{-2v|\tau|}(e^{j\omega\tau} + e^{-j\omega\tau})\,d\tau$$

or
$$S(f) = 2V^2 \int_0^\infty e^{-2v|\tau|} \cos \omega\tau \, d\tau$$

which is a standard integral of the form

$$\int e^{\alpha x} \cos \beta x \, dx = \frac{e^{\alpha x}}{\alpha^2 + \beta^2} [\alpha \cos \beta x + \beta \sin \beta x]$$

Hence, we obtain

$$S(f) = 2V^2 \left[\frac{e^{-2v|\tau|}}{(2v)^2 + \omega^2} \{ -2v \cos \omega\tau + \omega \sin \omega\tau \} \right]_0^\infty$$

$$= \frac{2V^2}{(2v)^2 + \omega^2} \{ 0 - e^0(-2v) \}$$

or
$$S(f) = \frac{4V^2 v}{(2v)^2 + \omega^2}$$

and the maximum value is $S(0) = V^2/v$. It is illustrated in Fig. 3.6.

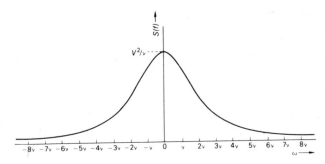

Fig. 3.6

The physical significance of the result is that Fig. 3.6 shows there is a fair amount of high-frequency power in the signal even at values of $\omega \geqslant 4v$.

Example 3.7

The autocorrelation function of an *RC* network is given by

$$R(\tau) = (S_0\omega_0/4)e^{-\omega_0|\tau|}$$

where S_0 is the power spectral density ($\omega > 0$) and $\omega_0 = 1/RC$. Hence, determine the rms output voltage of the network.

Solution

From the Wiener–Khintchine theorem we have

$$R(\tau) = \int_{-\infty}^{+\infty} S(f)e^{j\omega\tau}\,df$$

or

$$R(0) = \int_{-\infty}^{+\infty} S(f)\,df$$

which is the average output power of the network. Here

$$R(0) = \frac{S_0\omega_0}{4}e^0 = \frac{S_0\omega_0}{4} \quad \text{watts}$$

If v_{rms} is the output voltage in a $1\,\Omega$ load then

$$P_{av} = v_{rms}^2$$

or

$$v_{rms} = \sqrt{\frac{S_0\omega_0}{4}} = \frac{\sqrt{S_0\omega_0}}{2} \quad \text{volts}$$

3.6 White noise

Random noise signals generated by various sources are known to have a uniform power distribution over a very wide range of frequencies up to about 10^{13} Hz, which is in the ultra-violet region, after which it falls off as predicted by quantum theory. Such noise is defined as *white noise* by analogy with white light which has a uniform power distribution over the band of optical frequencies. The nearest examples are thermal noise in a resistor and shot noise in valves or transistors which also have a Gaussian amplitude distribution and are known as *Gaussian white noise*.

If N_0 is the noise power spectral density per Hz for *positive* frequencies only then, assuming both positive and negative frequencies (for mathematical purposes), the noise power spectral density is $N_0/2$, as shown in Fig. 3.7(a). Hence, we obtain for the autocorrelation function

$$R(\tau) = \frac{1}{2\pi} \int_{-\infty}^{+\infty} S(\omega)e^{j\omega\tau}\,d\omega = \frac{1}{2\pi} \int_{-\infty}^{+\infty} \frac{N_0 e^{j\omega\tau}}{2}\,d\omega = \frac{N_0}{2}\frac{1}{2\pi} \int_{-\infty}^{+\infty} e^{j\omega\tau}\,d\omega$$

or

$$R(\tau) = \frac{N_0}{2}\delta(\tau)$$

where $\delta(\tau)$ is the Dirac delta function shown in Fig. 3.7(b). Since $R(\tau)$ has a value at $\tau = 0$ only, there is no correlation between any two samples of white noise separated by an interval $\tau > 0$ and they are therefore statistically independent.

From Fig. 3.7(a) it will be observed that the average power, which is given by

$$P_{av} = \frac{1}{2\pi} \int_{-\infty}^{+\infty} S(\omega)\,d\omega$$

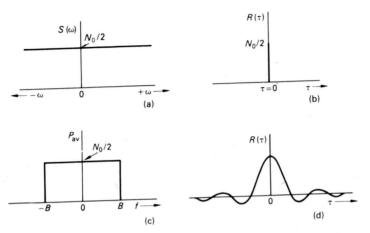

Fig. 3.7

becomes infinite and cannot be physically realised in any practical circuits. As most communication circuits are band-limited, it is more practical to consider the results of passing white noise through a filter with some defined bandwidth. The output noise is then called band-limited white noise or *coloured noise*.

3.7 Band-limited white noise

If white noise is passed through a low-pass ideal filter with a bandwidth $\pm B$ Hz, the output noise can be obtained by means of the transfer function $H(\omega)$ of the filter. Hence, we have

$$\frac{S_o(\omega)}{S_i(\omega)} = |H(\omega)|^2$$

where $S_i(\omega)$ and $S_o(\omega)$ are the input and output power spectral densities and $|H(\omega)| = 1$, with

$$P_{av} = \frac{1}{2\pi} \int_{-\infty}^{+\infty} S_o(\omega)\,d\omega = \frac{1}{2\pi} \int_{-\infty}^{+\infty} \frac{N_0}{2}\,d\omega = \frac{N_0}{2} \int_{-B}^{+B} df = N_0 B \quad \text{watts}$$

and is illustrated in Fig. 3.7(c).

The autocorrelation function $R(\tau)$ of the filtered white noise is

$$R(\tau) = \frac{1}{2\pi} \int_{-\infty}^{+\infty} S_o(\omega)e^{j\omega\tau}\,d\omega = \frac{1}{2\pi} \int_{-\infty}^{+\infty} S_i(\omega)\,|H(\omega)|^2\,e^{j\omega\tau}\,d\omega$$

$$= \frac{1}{2\pi} \int_{-\infty}^{+\infty} \frac{N_0}{2} e^{j\omega\tau}\,d\omega = \frac{N_0}{2} \int_{-B}^{+B} e^{j2\pi f\tau}\,df = \frac{N_0}{2}\left[\frac{e^{j2\pi f\tau}}{j2\pi\tau}\right]_{-B}^{+B}$$

$$= N_0 B \frac{\sin(2\pi B\tau)}{2\pi B\tau}$$

or $$R(\tau) = N_0 B \frac{\sin x}{x}$$

where $x = 2\pi B\tau$ and is shown in Fig. 3.7(d).

Since the shape of $R(\tau)$ is a $(\sin x)/x$ function, there is correlation on either side of $\tau = 0$ and the correlation is periodic at intervals of $\tau = 1/2B$. Hence, filtering uncorrelated white noise produces *correlated*, band-limited white noise or coloured noise.

Example 3.8

The input to an *RC* low-pass filter shown in Fig. 3.8 is white noise with a power spectral density of $(N_0/2)$ watts/Hz. Determine the output power spectral density and average noise power.

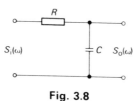

Fig. 3.8

Solution

If $H(\omega)$ is the network transfer function we have

$$H(\omega) = \frac{1/j\omega C}{R + 1/j\omega C} = \frac{1}{1 + j\omega RC}$$

or $$|H(\omega)|^2 = \frac{1}{1 + (\omega RC)^2}$$

If $S_i(\omega)$ and $S_o(\omega)$ are the input and output power spectral densities respectively then

$$\frac{S_o(\omega)}{S_i(\omega)} = |H(\omega)|^2 = \frac{1}{1 + (\omega RC)^2}$$

or $$S_o(\omega) = \frac{N_0/2}{1 + (\omega RC)^2}$$

The average output noise power is given by

$$P_{av} = \frac{1}{2\pi} \int_{-\infty}^{+\infty} S_o(\omega)\, d\omega = \frac{N_0}{4\pi} \int_{-\infty}^{+\infty} \frac{d\omega}{1 + (\omega RC)^2}$$

Using the substitution $u = \omega RC$ with $du = RC\, d\omega$ we obtain

$$P_{av} = \frac{N_0}{4\pi RC} \int_{-\infty}^{+\infty} \frac{du}{u^2+1} = \frac{N_0}{4\pi RC} \left[\tan^{-1} u\right]_{-\infty}^{+\infty}$$

$$= \frac{N_0}{4\pi RC}\left[\tan^{-1}\infty - \tan^{-1}(-\infty)\right]$$

$$= \frac{N_0}{4\pi RC}\left[\frac{\pi}{2} - \left(-\frac{\pi}{2}\right)\right]$$

or

$$P_{av} = N_0/4RC$$

Example 3.9

A sine-wave carrier together with Gaussian white noise is passed through a narrowband IF amplifier and detected by a linear envelope detector. Show that the output signal follows either a Rayleigh distribution or a Gaussian distribution depending on the input carrier-to-noise ratio.

Solution
If the carrier wave is $v = V_c \sin \omega_c t$ then the carrier plus band-limited noise from the IF amplifier $v(t)$ is given by

$$v(t) = V_c \sin \omega_c t + n(t) = \{V_c + x(t)\} \sin \omega_c t + y(t) \cos \omega_c t$$

where the noise signal $n(t)$ is expressed by its in-phase and quadrature phase components. (See Appendix E.)
Substituting $x = V_c + x(t)$ and $y = y(t)$ then yields

$$v(t) = x \sin \omega_c t + y \cos \omega_c t$$

or

$$v(t) = r \sin[\omega_c t + \phi]$$

where $r = \sqrt{x^2 + y^2}$ and $\phi = \tan^{-1}(y/x)$.
The variables x and y can be shown to be *independent* Gaussian distributions such that

$$p(x) = \frac{e^{-(x-V_c)^2/2\sigma^2}}{\sqrt{2\pi\sigma^2}} \quad \text{and} \quad p(y) = \frac{e^{-y^2/2\sigma^2}}{\sqrt{2\pi\sigma^2}}$$

where σ^2 is the variance.
The distributions of the amplitude r and phase angle ϕ will depend jointly on x and y. Since variables x and y transform directly into variables r and ϕ, we have

$$p(x, y)\, dx\, dy = p(r, \phi)\, dr\, d\phi$$

Also

$$p(x, y) = p(x)p(y) = \frac{e^{-[(x-V_c)^2 + y^2]/2\sigma^2}}{2\pi\sigma^2}$$

$$= \frac{e^{-[(x^2+y^2) - 2xV_c + V_c^2]/2\sigma^2}}{2\pi\sigma^2}$$

or
$$p(x, y)\,dx\,dy = \frac{e^{-[r^2 - 2rV_c\cos\phi + V_c^2]/2\sigma^2}}{2\pi\sigma^2}\,r\,dr\,d\phi$$

since $dx\,dy = r\,dr\,d\phi$. Hence

$$p(r, \phi) = \frac{re^{-[r^2 - 2rV_c\cos\phi + V_c^2]/2\sigma^2}}{2\pi\sigma^2}$$

To obtain the amplitude distribution $p(r)$ we must integrate this expression over all values of ϕ from 0 to 2π. Hence

$$\begin{aligned}
p(r) &= \int_0^{2\pi} p(r, \phi)\,d\phi \\
&= \int_0^{2\pi} \frac{re^{-[r^2 - 2rV_c\cos\phi + V_c^2]/2\sigma^2}}{2\pi\sigma^2}\,d\phi \\
&= \frac{re^{-(r^2 + V_c^2)/2\sigma^2}}{2\pi\sigma^2} \int_0^{2\pi} e^{rV_c\cos\phi/\sigma^2}\,d\phi
\end{aligned}$$

To evaluate this integral we use the standard integral $I_0(z)$ where

$$I_0(z) = \frac{1}{2\pi} \int_0^{2\pi} e^{z\cos\phi}\,d\phi$$

which is a modified Bessel function of zero order. Hence

$$p(r) = \frac{re^{-(r^2 + V_c^2)/2\sigma^2}}{\sigma^2} I_0\left(\frac{rV_c}{\sigma^2}\right)$$

The quantity V_c/σ^2 can be related to the input carrier and noise powers C_i and N_i respectively since $V_c^2/2 = C_i$ and $\sigma^2 = \overline{n^2(t)} = N_i$. Hence

$$\frac{rV_c}{\sigma^2} = \frac{r\sqrt{2C_i}}{\sigma\sqrt{N_i}}$$

or
$$z = \frac{r}{\sigma}\sqrt{2(C_i/N_i)}$$

When $C_i \ll N_i (z \ll 1)$ then $I_0(z) \simeq e^{z^2/4} \simeq 1$ and we obtain

$$p(r) = \frac{re^{-(r^2 + V_c^2)/2\sigma^2}}{\sigma^2}$$

or
$$p(r) \simeq \frac{re^{-r^2/2\sigma^2}}{\sigma^2} \qquad \text{(since } V_c^2/2\sigma^2 \simeq 0\text{)}$$

which is a Rayleigh distribution with the peak value at $r = \sigma$.

When $C_i \gg N_i$ ($z \gg 1$) then we have $I_0(z) \simeq e^z/\sqrt{2\pi z}$. Hence

$$p(r) = \frac{re^{-(r^2 + V_c^2)/2\sigma^2}}{\sigma^2} \frac{e^{rV_c/\sigma^2}}{\sqrt{2\pi r V_c/\sigma^2}}$$

$$= \frac{re^{-(r - V_c)^2/2\sigma^2}}{\sqrt{2\pi r V_c \sigma^2}}$$

or $$p(r) \simeq \frac{e^{-(r - V_c)^2/2\sigma^2}}{\sigma\sqrt{2\pi}} \qquad (\text{near } r \simeq V_c)$$

which is a Gaussian distribution centred at $r = V_c$. The distributions are shown in Fig. 3.9.

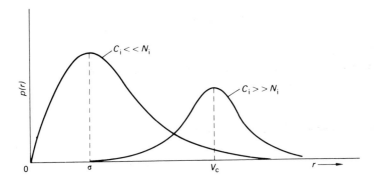

Fig. 3.9

Comment
In the case when *only* random noise enters the receiver, the analysis is similar to that for $C_i \ll N_i$ with $V_c = 0$. Hence, the noise output from the detector will have a Rayleigh distribution as obtained previously.

3.8 Correlation detection[11, 15]

The detection of a weak periodic signal which is embedded in random white noise may be achieved by using autocorrelation or cross-correlation techniques. If the signal is autocorrelated, a correlation function is obtained as shown in Fig. 3.10(a).

The large peak at zero delay τ is caused by the autocorrelation of the noise and the autocorrelation of the periodic component. For large values of delay τ, the autocorrelation function of the random signal with zero mean tends to zero,

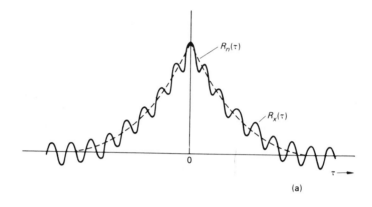

(a)

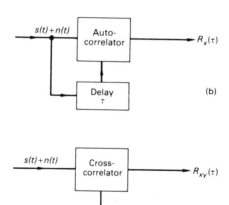

(b)

(c)

Fig. 3.10

while that of the periodic signal is still periodic. Hence, the periodic signal can be detected and its frequency determined.

Assuming the periodic signal $s(t)$ and the random noise signal $n(t)$ are uncorrelated, for $x(t) = s(t) + n(t)$ we have

$$R_x(\tau) = R_s(\tau) + R_{sn}(\tau) + R_{ns}(\tau) + R_n(\tau)$$

where $R_{sn}(\tau) = R_{ns}(\tau) = 0$ since the periodic signal and noise are uncorrelated, while $R_n(\tau) \to 0$ for large values of τ. Hence, we obtain

$$R_x(\tau) = R_s(\tau)$$

and the technique is illustrated in Fig. 3.10(b).

Since autocorrelation does not reveal information about the waveform of $s(t)$, this can be obtained by cross-correlation. The signal $s(t)$ is cross-correlated with an impulse signal $y(t)$ whose periodic time T_0 is the same as that of $s(t)$ and may be determined by autocorrelation (as shown above) if it is not known. Hence, if $n(t)$ and $y(t)$ are uncorrelated we have

$$R_{xy}(\tau) = \lim_{T \to \infty} \frac{1}{T} \int_{-T/2}^{+T/2} [s(t) + n(t)] \, y(t - \tau) \, dt$$

or

$$R_{xy}(\tau) = \lim_{T \to \infty} \frac{1}{T} \left[\int_{-T/2}^{+T/2} s(t) y(t - \tau) \, dt + \int_{-T/2}^{+T/2} n(t) y(t - \tau) \, dt \right]$$

The second term on the r.h.s. is the cross-correlation function $R_{ny}(\tau)$ which is zero as the signal and noise are uncorrelated. Hence, we obtain

$$R_{xy}(\tau) = \frac{1}{T_0} \int_{-T_0/2}^{+T_0/2} s(t) \, y(t - \tau) \, dt$$

where T_0 replaces T as averaging over T is the same as over any period T_0, since $s(t)$ and $y(t)$ have the same periodic time T_0. The technique is similar to the sampling principle and it can be shown that

$$R_{xy}(\tau) = \frac{1}{T_0} s(t)$$

for all values of τ. The output is an exact replica of the transmitted signal and requires only a knowledge of the transmitted frequency. It is illustrated in Fig. 3.10(c).

Signal detection is therefore achieved by using an autocorrelator or cross-correlator as is appropriate. It will be shown in the next section that correlation detection in the time domain is similar to matched filtering in the frequency domain. Furthermore, correlation techniques can also be used for non-periodic signals in random white noise, such as a pseudo-random coded signal. Correlators usually employ product multipliers or phase-sensitive detectors which are followed by an averaging circuit.

3.9 Matched filter

In the filtering of analogue signals, a linear filter is specified and the output is tacitly accepted as such filtering discriminates against large frequency errors but does so at the expense of small frequency deviations. However, in the filtering of digital signals, it is pointless having an output *anywhere* between 0 and 1. Hence, on the basis of the received waveform, it is necessary to decide which of the two states the signal is in.

In this decision approach, the filter at the receiver must be matched to the received waveform to achieve the maximum output signal-to-noise ratio. Such a filter is designed specifically to *maximise* the output signal-to-noise ratio and is called a *matched filter*. The matched filter is the optimum filter for detecting signals received with additive white noise, i.e. noise with a uniform power density spectrum over a wide frequency band.

It is shown in Appendix F that the transfer function of such a filter is given by

$$H(\omega) = kS^*(\omega)e^{-j\omega t_d}$$

where k is an arbitrary gain constant, $S^*(\omega)$ is the complex conjugate of $S(\omega)$, the spectrum of the received signal $s(t)$, and t_d is the time-delay of the filter. Furthermore, the impulse response of such a filter is given by

$$h(t) = ks(t_d - t)$$

where $s(t)$ is the input signal to which the filter is matched. The time-delay t_d is required to make the filter physically realisable and the filter has its maximum output at some time $t = t_0$. The optimum decision is then made at time $t = t_0$ to determine the nature of the output corresponding to a 0 or 1, and the output depends only on the original signal energy and is independent of its waveform. A practical implementation of the matched filter for a rectangular pulse is shown in Fig. 3.11.

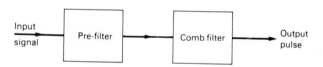

Fig. 3.11

In many practical cases, it is not feasible to provide a suitable matched filter as it may be physically unrealisable and so somewhat simpler filters are used with some loss in signal-to-noise ratios. Table 3.1 lists some typical examples of these and it is observed that the maximum loss is only about 1 to 2 dB.

In Appendix F it is shown that the matched filter is mathematically equivalent to a correlation detector. In practical applications it may therefore be more convenient to implement the matched filter by means of a correlation detector. In the correlation detector shown in Fig. 3.12(a), the transmitted signal waveform $s(t)$ is stored at the receiver and correlated (multiplied) with the received signal plus noise. After integration over the time interval t_0, the detector then decides whether a 0 or 1 has been received.

Table 3.1

Signal	Filter	Loss (dB)
Rectangular pulse	Rectangular	1·7
Rectangular pulse	Gaussian	0·98
Gaussian pulse	Rectangular	0·98
Gaussian pulse	Gaussian	0

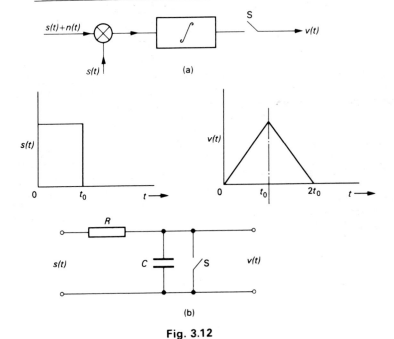

Fig. 3.12

A simple practical realisation of the matched filter for a rectangular waveform is the integrate and dump filter shown in Fig. 3.12(b). If the input rectangular pulse is convolved with the impulse response of the matched filter, it yields a triangular output waveform. This linear rise in voltage can be obtained by applying the rectangular pulse to an RC network with $RC \gg t_0$. At time t_0, the output level is sampled and then shorted momentarily to reset the filter to its initial value.

Pre-whitening filter

In the case of non-white or 'coloured' noise, it is found that optimum detection by a matched filter is best obtained by preceding it with a pre-whitening filter which converts the non-white noise to true white noise and modifies the input signal slightly. The modified signal is now mixed with white noise and can be optimally detected as was shown earlier. The modification to the input signal leads to some intersymbol interference in the case of digital signals which can be minimised if the bandwidth of the pre-whitening filter is large compared to $1/T$ where T is the duration of the digital signal.

3.10   Decision theory[4]

In the design of optimum receivers, an important problem is the detection of the received signal in a background of noise. This may involve simply the detection of a given signal or its absence or it may involve the detection of a zero or one, as in digital data communications. In either case, the receiver must make the best possible decision on the basis of various criteria. The statistical nature of the decision making process involves hypothesis testing and is known as *decision theory* or *detection theory*.

Receivers designed to minimise the average cost of making a decision use, as a basis, Bayes' decision rule. When a sufficient knowledge of costs and *a priori* probabilities is available, the decision process can be optimised according to an expected cost criterion.

It is shown in Appendix G that such decision procedures which are based on expected cost minimisation involve testing a *likelihood ratio* which is the ratio of the *a posteriori* probabilities of the observations against a threshold which depends on the *a priori* probabilities and costs. It is given by the expression

$$L[y(t)] \underset{H_1}{\overset{H_2}{\gtrless}} \frac{P_1(C_{21}-C_{11})}{P_2(C_{12}-C_{22})}$$

where the r.h.s. is denoted by the threshold value L_t, P_1 and P_2 are the *a priori* probabilities associated with the hypotheses H_1 and H_2 respectively, and C_{11}, C_{12}, C_{21}, and C_{22} are the conditional costs of making decisions.

The Bayes' criterion is therefore characterised by the average cost or risk involved in making a decision. In many cases, it is more useful to express the risk in terms of the probabilities of detection P_D and false alarm P_F, provided the *a priori* probabilities and costs are available.

If the costs and *a priori* probabilities are not available, a useful decision strategy is the *Neyman–Pearson criterion* which maximises P_D while holding P_F at some acceptable value. This type of criterion can also be reduced to a likelihood ratio, where the threshold is determined by the allowed false alarm probability. It is shown in Appendix G that its value is given by

$$L[y(t)] = \frac{dP_D}{dP_F}$$

where (dP_D/dP_F) is the slope of the receiver operating characteristic at any given point.

3.11 Estimation theory[16]

It was shown earlier that decision theory can be used to *detect* the presence of a signal. Similarly, *estimation theory* can be used to make an *estimate* of some unknown parameter of a signal such as its frequency or phase. Since the signal is received with additive noise, only an estimate of the signal parameter is possible. Two useful estimators are Bayes' estimate, which endeavours to *minimise* a cost function, and the maximum-likelihood estimate, which tends to *maximise* a likelihood function. As the latter does not require any *a priori* information, it is often employed and will be considered here.

Maximum-likelihood estimation (MLE)
Since the received signal is a function of time and contains additive noise, it can be represented by

$$z(t) = s(t, \theta) + n(t)$$

where $s(t, \theta)$ is the transmitted signal, θ is the unknown parameter to be estimated, and $n(t)$ is assumed to be Gaussian white noise. By observing the signal continuously or by sampling it, it is possible to make an estimate of θ which is denoted by $\hat{\theta}$.

It is shown in Appendix H that, for a sampled signal, the likelihood function of $z(t)$ is

$$p(z|\theta) = A \exp \left\{ -\frac{1}{N_0} \int_0^T [z(t) - s(t, \theta)]^2 \, dt \right\}$$

where A is an unknown constant. To obtain the maximum-likelihood function, it is convenient to differentiate $\ln p(z|\theta)$ with respect to θ and then equate it to zero. This yields

$$\frac{\partial}{\partial \theta} [\ln p(z|\theta)] = \frac{2A}{N_0} \int_0^T [z(t) - s(t, \theta)] \frac{\partial s(t, \theta)}{\partial \theta} dt$$

and so the maximum-likelihood estimate of θ is a solution of the equation

$$\int_0^T [z(t) - s(t, \hat{\theta})] \frac{\partial s(t, \hat{\theta})}{\partial \theta} \, dt = 0$$

where θ has been replaced by the estimate $\hat{\theta}$.

The estimate $\hat{\theta}$ depends on the signal received and on the number of observations made. Hence, it can vary as a random variable with a mean and variance. The variance of an estimator, under certain conditions, cannot be less than the lower bound known as the Cramér–Rao bound.

It is shown in Appendix H that, if the expectation of $\hat{\theta}$, i.e. $E[\hat{\theta}|\theta]$, is equal to the true value θ, the variance of $\hat{\theta}$ is given by

$$\sigma_{\hat{\theta}}^2 \geqslant \frac{1}{E\left[\left\{ \dfrac{\partial \ln p(z|\theta)}{\partial \theta} \right\}^2 \right]}$$

Example 3.10

A signal of known amplitude and frequency has the form $s(t, \theta) = A \sin(\omega_0 t + \phi)$ where $0 \leqslant t \leqslant T$. Make an estimate of the phase angle ϕ.

Solution

We have

$$s(t, \theta) = A \sin(\omega_0 t + \phi)$$

$$\frac{\partial s(t, \theta)}{\partial \theta} = A \cos(\omega_0 t + \phi)$$

and the phase estimate $\hat{\phi}$ is a solution of the equation

$$\int_0^T [z(t) - A \sin(\omega_0 t + \hat{\phi})] A \cos(\omega_0 t + \hat{\phi}) \, dt = 0$$

or

$$\int_0^T z(t) \cos(\omega_0 t + \hat{\phi}) \, dt = \int_0^T \frac{A}{2} \sin 2(\omega_0 t + \hat{\phi}) \, dt$$

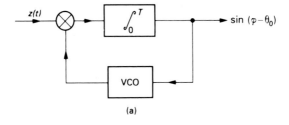

(a)

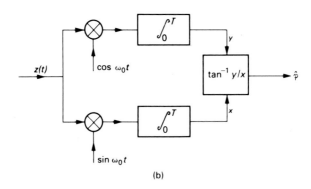

(b)

Fig. 3.13

The integral on the r.h.s. is zero if $\omega_0 T = n\pi$ where n is an integer or $\omega_0 T \gg 1$. Hence

$$\int_0^T z(t)\cos(\omega_0 t + \hat{\phi})\,dt = 0$$

and $\hat{\phi}$ can be determined by using the phase-locked loop shown in Fig. 3.13(a). When the loop is locked, the phase angle θ_0 of the VCO nearly equals that of $z(t)$, i.e. $\theta_0 \simeq \phi$.

Alternatively, by expanding the cosine term and rearranging the expression we obtain

$$\sin\hat{\phi}\int_0^T z(t)\sin\omega_0 t\,dt = \cos\hat{\phi}\int_0^T z(t)\cos\omega_0 t\,dt$$

or

$$\hat{\phi} = \tan^{-1}\left[\frac{\displaystyle\int_0^T z(t)\cos\omega_0 t\,dt}{\displaystyle\int_0^T z(t)\sin\omega_0 t\,dt}\right]$$

and $\hat{\phi}$ can be determined by using two multipliers and matched filters as shown in Fig. 3.13(b).

4

Circuit noise

The two most important types of noise associated with electronic components, such as valves, transistors, and resistors, are *thermal* noise and *shot* noise. The physical basis of each will be considered and, as both of them give rise to noise power in the system, they can be regarded as producing one combined noise effect.

4.1 Thermal noise

A metallic conductor or resistor contains a number of free electrons. Due to thermal agitation, these free electrons are moving about continuously in the conductor causing collisions with the atoms and a continuous exchange of energy takes place. This accounts for the resistance property of the conductor and, though there is no current in the conductor on open-circuit, the random motion of electrons in the conductor produces voltage fluctuations across the conductor which accounts for a mean-square noise voltage $\overline{v_n^2}$ at its terminals.

The thermal noise effect was investigated experimentally by Johnson[17] and theoretically by Nyquist.[18] Experimental results showed that the thermal noise voltage depends upon temperature and its mean-square value $\overline{v_t^2}$ is given by

$$\overline{v_t^2} = 4kTBR$$

where k is Boltzmann's constant, T is the absolute temperature, B is the bandwidth of the system, and R is the resistance of the conductor. For example, if $R = 1\,\mathrm{k\Omega}$, $B = 5\,\mathrm{MHz}$, and $T = 290\,\mathrm{K}$ then with $k = 1.38 \times 10^{-23}\,\mathrm{J/K}$ we obtain $\overline{v_t^2} = 80 \times 10^{-12}$ or $v_{\mathrm{rms}} = [\overline{v_t^2}]^{1/2} \simeq 9\,\mu\mathrm{V}$.

Nyquist's investigation of the effect was based on thermodynamical reasoning and similar results were obtained. He showed that the thermal noise power P_n associated with any resistor is given by

$$P_n = kTB \quad \text{watts}$$

where k, T, and B have their previous meaning.

The derivation is given in Appendix I and is based on the assumption of *available noise power*. This implies matched conditions, as is usually the case in most communication channels, since it is necessary to transfer the maximum signal power through the system. However, in practice, the concepts of noise voltage, noise power, or noise power spectral density can be equally well employed in the study of noise problems. If the noise *voltage* spectral density is $S_v(f)$, it can be shown that*

$$S_v(f) = 2kTR$$

a result which depends on T and R but is independent of frequency up to about 10^{13} Hz. This implies that thermal noise covers a broad band of frequencies and has a uniform response. Hence, it is often called 'Johnson noise' or *white noise* due to an analogy with white light which has a uniform power distribution over the band of optical frequencies.

Equivalent circuit
It is convenient in practice to represent thermal noise in a resistor as due to a thermal noise source $\overline{v_t^2}$ in series with a noiseless resistor R, which is based on Thévenin's theorem. Alternatively, a current source $\overline{i_t^2}$ in shunt with a conductance G may be used and this is based on Norton's theorem. This is illustrated in Fig. 4.1.

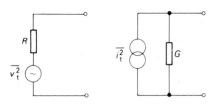

Fig. 4.1

Under matched conditions, the load is also R (assumed noiseless) and the maximum noise power available from the source is obtained as follows.

In Fig. 4.2, if i is the current in the circuit then $i = \sqrt{\overline{v_t^2}}/2R$ and the maximum

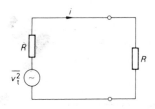

Fig. 4.2

* See Appendix C.

noise power delivered to the load is P_{max} where

$$P_{\text{max}} = i^2 R = \overline{v_t^2}/4R^2 \times R = \overline{v_t^2}/4R$$

From Nyquist's results, $P_{\text{max}} = P_n = kTB$ and we obtain

$$\overline{v_t^2}/4R = kTB$$

or
$$\overline{v_t^2} = 4kTBR$$

a result which was verified, experimentally, by Johnson.

For the current source, likewise, we have

$$\overline{i_t^2} = G^2 \overline{v_t^2} = 4kTBG$$

Comment
In the design of RF amplifiers, the input resistance R_g should have a low value of about $50\,\Omega$ as noise generated at the input is amplified in subsequent stages. The choice of $50\,\Omega$ is also concerned with maximum signal power transfer and matching.

Example 4.1
Two resistors R_1 and R_2 are connected (a) in series and (b) in parallel. If the temperatures of the resistors in equilibrium are T_1 and T_2 respectively, calculate the rms noise voltage at the output terminals in each case.

Solution
(a)   Series resistors

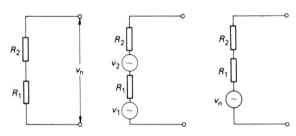

Fig. 4.3

If v_n is the rms noise voltage at the output terminals then

$$v_n^2 = v_1^2 + v_2^2$$

where
$$v_1^2 = 4kBT_1 R_1$$
$$v_2^2 = 4kBT_2 R_2$$

Hence
$$v_n^2 = 4kBT_1 R_1 + 4kBT_2 R_2$$

or
$$v_n = \sqrt{4kB(R_1 T_1 + R_2 T_2)}$$

In particular, if $T_1 = T_2 = T$, as is generally the case in practice, then

$$v_n = \sqrt{4kTB(R_1 + R_2)}$$

and the two resistors behave as a *single* noisy resistor of value $(R_1 + R_2)$.

(b) Parallel resistors

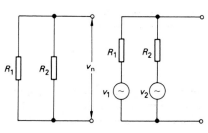

Fig. 4.4

The noise voltage at the output terminals due to v_1 only is $v_{o_1} = v_1 R_2/(R_1 + R_2)$ and the noise voltage at the output terminals due to v_2 only is $v_{o_2} = v_2 R_1/(R_1 + R_2)$. Since these two voltages are rms voltages, the total rms voltage at the output terminals is v_n where $v_n = \sqrt{v_{o_1}^2 + v_{o_2}^2}$. Hence, we have

$$v_1^2 = 4kBT_1 R_1$$

$$v_2^2 = 4kBT_2 R_2$$

with
$$v_{o_1}^2 = \frac{4kBT_1 R_1 R_2^2}{(R_1 + R_2)^2}$$

$$v_{o_2}^2 = \frac{4kBT_2 R_1^2 R_2}{(R_1 + R_2)^2}$$

and
$$v_n^2 = \frac{4kBR_1 R_2 (T_1 R_2 + T_2 R_1)}{(R_1 + R_2)^2}$$

or
$$v_n = \sqrt{\frac{4kBR_1 R_2}{(R_1 + R_2)^2}(T_1 R_2 + T_2 R_1)}$$

In particular, when $T_1 = T_2 = T$, as is usually the case in practice, then

$$v_n = \sqrt{\frac{4kBR_1 R_2 T}{(R_1 + R_2)^2}(R_1 + R_2)}$$

or
$$v_n = \sqrt{\frac{4kTBR_1 R_2}{(R_1 + R_2)}}$$

and the two resistors behave as a single noisy resistor of value

$$\frac{R_1 R_2}{(R_1 + R_2)}$$

4.2   Shot noise[19]

The current flow in a vacuum diode is due to the emission of electrons from the cathode which then travel to the anode. Each electron carries a discrete amount of charge to the anode and produces a small current pulse. The summation of all the current pulses produces the average anode current I_a in the diode.

However, the emission of electrons is a random process depending on the surface condition of the cathode, shape of the electrodes, and the potential between them. This gives rise to random fluctuations in the number of emitted electrons and so the diode current contains a time-varying component. Since each electron arriving at the anode is like a 'shot', the fluctuating component gives rise to a mean-square shot noise current $\overline{i_s^2}$.

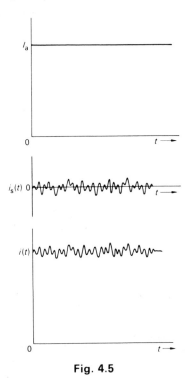

Fig. 4.5

The fluctuations are most pronounced in the temperature-limited diode and so the shot noise component is greatest. Hence, if the instantaneous diode current is $i(t)$ it can be written as

$$i(t) = I_a + i_s(t)$$

where I_a is the average anode current and $i_s(t)$ is the shot noise component. This is illustrated in Fig. 4.5.

From statistical considerations, it can be shown that the electron emission corresponds to a Poisson distribution from which it is possible to obtain an expression for the shot noise mean-square current $\overline{i_s^2}$. An alternative but simpler proof using the idea of energy spectral density is given in Appendix J.

If I_a is the average anode current, B is the effective bandwidth of the system, and e is the electronic charge, then the rms value of shot noise current is given by I_s where

$$I_s = \sqrt{2eI_a B}$$

for a temperature-limited diode. The equivalent circuit is shown in Fig. 4.6 where $\overline{i_s^2}$ is the equivalent current generator and the diode anode resistance r_a is assumed to be infinite and is therefore omitted from the equivalent circuit.

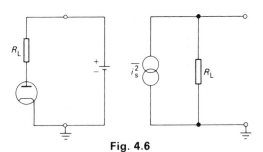

Fig. 4.6

In the case of a diode or triode, operating under space-charge conditions, the presence of the space-charge produces a smoothing effect on the random fluctuations and so the rms shot noise current I_s is smaller. This is accounted for by using a factor k where $0 < k < 1$ with

$$I_s^2 = 2k^2 eI_a B$$

As r_a for a triode is finite, it must be considered and the equivalent circuit is given in Fig. 4.7.

Equivalent noise resistance
The effect of shot noise is to produce further noise in the circuit, as does thermal noise, and so it can be reduced to 'equivalent' thermal noise by associating it

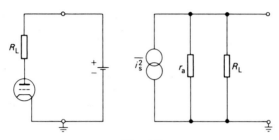

Fig. 4.7

with an equivalent resistor R_{eq}. For a triode, the hypothetical resistor is inserted in series with the grid and it is given by

$$R_{eq} \simeq 2 \cdot 5/g_m$$

where g_m is the mutual conductance of the triode. The equivalent circuit for combined thermal and shot noise is shown in Fig. 4.8.

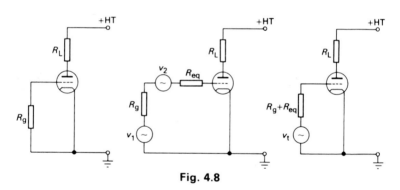

Fig. 4.8

The equivalent rms 'thermal' noise voltage v_t of Fig. 4.8 is given by

$$v_t = \sqrt{v_1^2 + v_2^2}$$

where

$$v_1^2 = 4kTBR_g$$

$$v_2^2 = 4kTBR_{eq}$$

Hence

$$v_t = \sqrt{4kTB(R_g + R_{eq})}$$

Example 4.2
State two sources of noise encountered in high-gain amplifier circuits and briefly explain their origins.

The input impedance of a high-gain amplifier is equivalent to a resistance of 4 kΩ over its operating range. A signal generator is connected to the amplifier by a transformer which matches the internal resistance of the generator to the amplifier. With the generator set to give zero signal output, the noise power is measured at the output terminals of the amplifier. If the output noise power decreases by 4 dB when the secondary terminals of the input transformer are short-circuited, calculate the equivalent grid noise resistance of the first valve.

Determine the noise factor of the arrangement under normal conditions (i.e. with the short-circuit removed from the transformer). (U.L.)

Solution

Assuming a generator impedance of 1 kΩ, the matching transformer turns ratio required is 1:2. The circuit is shown in Fig. 4.9(a) and the equivalent circuit (referred to the secondary) in Fig. 4.9(b).

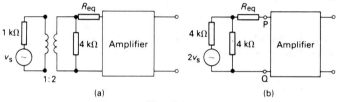

Fig. 4.9

If R_{eq} is the equivalent amplifier noise resistance, let P_1 be the noise power when the transformer is correctly terminated and P_2 be the noise power when the transformer is short-circuited. Hence, we have

$$10 \log \frac{P_1}{P_2} = 4$$

or

$$P_1/P_2 = 2 \cdot 5$$

Also

$$P_1 \propto v_1^2 \quad \text{with} \quad v_1^2 = 4kTB \left(\frac{4000}{2} + R_{eq} \right)$$

$$P_2 \propto v_2^2 \quad \text{with} \quad v_2^2 = 4kTBR_{eq}$$

or

$$\frac{4kTB(2000 + R_{eq})}{4kTBR_{eq}} = 2 \cdot 5$$

or

$$2 \cdot 5 R_{eq} = 2000 + R_{eq}$$

and

$$R_{eq} = \frac{2000}{1 \cdot 5} = 1 \cdot 33 \, \text{k}\Omega$$

The noise factor of the amplifier is defined in Section 5.1 as

$$F = \frac{(S_i/N_i)}{(S_o/N_o)}$$

where (S_i/N_i) is defined at the *source* and (S_o/N_o) may be referred to the input side at points P and Q, with the amplifier noise represented by R_{eq}. Hence

$$F = \frac{(2v_s)^2}{4000kTB} \bigg/ \frac{(v_s)^2}{(2000 + 1330)kTB}$$

$$= \frac{4}{4000} \times 3330$$

or $\qquad\qquad F = 3\cdot 33 \text{ or } 5\cdot 2\,\text{dB}$

4.3   Partition noise

In multigrid valves, such as tetrodes and pentodes, the division of current to one or other electrode is subject to random fluctuations also. This gives rise to a further noise effect which is basically similar to shot noise. It is known as *partition noise* and can be evaluated using statistical ideas similar to that for calculating shot noise.

This effect can be accounted for by increasing the value of the equivalent noise resistance R_{eq} obtained previously for the case of a triode. The value of R_{eq} for a pentode is given by

$$R_{eq} = \frac{I_a}{(I_a + I_s)}\left[\frac{2\cdot 5}{g_m} + \frac{20I_s}{g_m^2}\right]$$

where I_a is the anode current, I_s is the screen current, and g_m is the mutual conductance of the valve. Typical values for R_{eq} are between $1000\,\Omega$ to $10\,\text{k}\Omega$.

Due to partition noise, multigrid valves are more noisy than triodes and should be avoided in the early stages of an amplifier if noise is of primary concern, as in low-noise amplifiers for space communications. The noise generated in the early stages is amplified in subsequent stages and will give a large noise output. In the case of multigrid mixers, the conversion conductance g_c is used instead of g_m for obtaining the value of R_{eq}. Since g_c is much smaller than g_m, values of R_{eq} around $100\,\text{k}\Omega$ are possible and so multigrid mixers are quite noisy.

4.4   Bipolar transistor noise[20,21]

Noise in junction transistors shows some similarity to that in valves and the three types of valve noise, namely thermal noise, shot noise, and partition noise, are found to exist in transistors. This is basically because random fluctuations in the movement of the charge carriers (electrons and holes) cause variations in the various transistor currents. However, their exact nature and evaluation is more complicated and still not clearly understood.

For the bipolar transistor, it is found that thermal noise is associated with the base region which produces an rms noise voltage V_b given by

$$V_b = \sqrt{4kTBr_b}$$

where T is the absolute temperature, r_b is the base resistance, and B is the effective bandwidth.

Shot noise is also produced in the base region due to the random fluctuations of minority charge carriers across the emitter–base junction, which causes the emitter current to fluctuate. This gives rise to a mean-square shot noise current and the rms value is given by

$$I_s = \sqrt{2eI_eB}$$

where e is the electronic charge, I_e is the d.c. emitter current, and B is the effective bandwidth.

Since recombination may also take place in the base region, fluctuations in the recombination of minority charge carriers in the base produces further shot noise, analogous to partition noise, which appears as fluctuations in the collector current and the rms shot noise current is given by

$$I_s = [2eI_c(1 - |\alpha|^2/\alpha_0)B]^{1/2}$$

where e and B have their previous meaning, I_c is the d.c. collector current, α_0 is the d.c. current-gain factor, and α is the current-gain factor at frequency f.

It is usual to represent the various noise effects in a transistor by an equivalent T-network comprising three noise generators. This is illustrated in Fig. 4.10 for the common-base circuit.

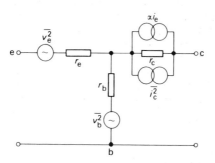

Fig. 4.10

In the circuit of Fig. 4.10, $\sqrt{\overline{v_b^2}}$ is the rms thermal noise component, $\sqrt{\overline{v_e^2}}$ is the rms shot noise component, and $\sqrt{\overline{i_c^2}}$ is the rms partition noise component. For practical purposes, it is convenient to evaluate noise in transistors in terms of a noise factor and details are given in Appendix K.

4.5 FET noise[22,23]

In both the JFET and MOSFET, the presence of a conducting channel gives rise to thermal noise which can be represented by an equivalent noise resistance R_n. For the JFET, this is given approximately by $R_n \simeq 0.65/g_m$ where g_m is the transconductance of the device at low frequencies. For the MOSFET, the value of R_n is somewhat higher and depends on the doping of the substrate and the thickness of the oxide layer under the gate.

At higher frequencies, due to capacitive coupling between the channel and gate, induced gate noise is produced which is partly correlated with the channel noise. For the JFET, the induced gate noise mean-square current is given by

$$\overline{i_g^2} = 4kTBK_i \omega^2 C_{gs}^2 / g_m$$

where the factor $K_i \simeq 0.3$ and C_{gs} is the gate-source input capacitance. For the MOSFET, the expression is similar but the factor $K_i \simeq 0.1$.

These two noise effects can be considered together and lead to the evaluation of a noise factor F as in the case of the bipolar transistor. It is shown in Appendix K that the minimum noise factor F_{min} for the JFET in the common-source or common-gate configuration is fairly similar and is given by

$$F_{min} \simeq 1 + 2R_n g_g + 2\sqrt{R_n g_g + R_n^2 g_g^2}$$

where R_n is the equivalent resistance of a noise voltage generator and g_g is the equivalent conductance of a noise current generator.

For the MOSFET, the minimum noise figure is given by

$$F_{min} \simeq 1 + 0.52\,\omega C/g_m$$

where C is the gate-to-channel capacitance and g_m is the forward transconductance.

Comments
1. At low frequencies, flicker noise or $1/f$ noise must also be considered. The effect is found to be greater in the MOSFET than in the JFET.
2. The common-gate and common-drain configurations have the same noise factor as the common-source configuration when feedback effects are neutralised.

5

Noise measurement

In many communication systems, the received signals are of low power and accompanied by noise. Further amplification is therefore necessary and this increases both signal and noise power levels. As the noise cannot be eliminated, it has been usual to use, as a criterion of performance, the ratio of signal power to noise power at various points in the system. This measure of system performance is not adequate in the case of certain active networks such as amplifiers or receivers. Consequently, it has been found necessary to use other criteria for measuring system performance.

5.1 Noise factor[24]

A familiar measure which is used extensively is the *noise factor* or *noise figure F*. In general, it measures the 'noisiness' of a network which may be an amplifier or receiver by considering two signal-to-noise ratios, one at the input end and the other at the output end of the network. This is illustrated in Fig. 5.1.

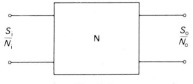

Fig. 5.1

If S_i/N_i is the ratio of signal power to noise power at the input of the network and S_o/N_o is the ratio of signal power to noise power at the output of the network, a measure of the effect of the network on the signal as it passes through the network can be obtained by comparing the two signal-to-noise ratios.

However, in defining the noise factor F, it is usual to introduce the concept of *maximum available power* or, more simply, the *available power*. Since it is observed in practice that the input impedance and source impedance, but not the output impedance, affect the value of F, it has been found useful to use the concept of available signal power and available noise power at the *signal source* or at the input if the source is connected directly to the input of the network. Hence, in terms of these quantities the noise factor F is defined as

$$F = \frac{(S_i/N_i)}{(S_o/N_o)}$$

where S_i = available input signal power
N_i = available input noise power
S_o = actual or available output signal power
N_o = actual or available output noise power

This definition implies that the input impedance of the network is matched to the signal source impedance, while the output load impedance may or may not be matched to the output network impedance. However, the matched condition usually applies in both cases in most communication systems, since matched conditions are generally used for various reasons. Nevertheless, in some special cases it may be more useful not to use matched impedances to obtain optimum conditions in the particular application.

For an input signal v_s with a source resistance R_s, the available signal power $S_i = v_s^2/4R_s$. For the case of thermal noise which produces a noise voltage v_n across the source resistance R_s, the available noise power is given by $v_n^2/4R_s = 4kTBR_s/4R_s = kTB$ as obtained by Nyquist. Hence

$$\frac{S_i}{N_i} = \frac{v_s^2}{4kTBR_s}$$

and the actual or available signal-to-noise ratio at the output is given by

$$\frac{S_o}{N_o} = \frac{v_s^2/4kTBR_s}{F} = \frac{v_s^2}{4kTBFR_s}$$

Since the definition of F above uses the signal-to-noise ratio at two different points in the system, it is sometimes more convenient to use an alternative expression obtained as follows.

$$F = \frac{S_i/N_i}{S_o/N_o} = \frac{N_o}{S_o/S_i \times N_i}$$

If G is the available power gain of the network then $G = S_o/S_i$ (matched input and output impedances) and we have also $N_i = kTB$. Hence

$$F = \frac{N_o}{GkTB} \quad \leftarrow \text{just } N_i \text{ amplified}$$

In the above expression, N_o is the noise power output of the network, while

$GkTB$ is the noise power output of an ideal network, i.e. a network which adds no further noise. Hence

$$F = \frac{N_\text{o}}{N_\text{o}'}$$

where $N_\text{o}' = GkTB$.

Comments
1. $F = 1$ for an ideal network because $N_\text{o} = N_\text{o}'$ with $F = N_\text{o}/N_\text{o}' = 1$.
2. T is usually taken as 290 K with $kT = 4 \times 10^{-21}$ W.
3. F may also be expressed in dB where $F_{\text{dB}} = 10 \log_{10} F$.
4. A typical value of F for a radar receiver is around 10 and for a low-noise amplifier (maser) around 1·1.

Example 5.1
Discuss the sources of noise in an amplifier and the manner in which they can limit amplifier performance.

A high-gain amplifier designed to operate from a 75 Ω source contains an input transformer having a step-up ratio of 1:4 preceding the first valve stage. Assuming the equivalent noise resistance of the valve to be 700 Ω, draw the equivalent circuit of the arrangement and hence calculate the source emf that would be required to yield unit S/N ratio at the output in a bandwidth of 200 kHz. The value of kT may be taken as $4\cdot14 \times 10^{-21}$ W.   (C.G.L.I.)

Solution
The main sources of noise arising in an amplifier are described in Sections 4.1 to 4.3. The overall effect of the noise is to reduce the output signal-to-noise ratio of the amplifier. If this ratio is below a certain minimum then the signal is of little value and the performance of the amplifier is limited.

Problem
The circuit is shown in Fig. 5.2(a) where e_s is the emf of the signal source and the equivalent circuit is given in Fig. 5.2(b). S_1/N_1 is the available signal-to-noise ratio at the

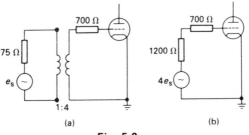

(a)   (b)

Fig. 5.2

input and S_o/N_o is the signal-to-noise ratio at the output, with $S_o/N_o = 1$. Also, we have

$$S_i = \frac{(4e_s)^2}{4 \times 1200} = \frac{e_s^2}{300}$$

$$N_i = kTB = 4 \cdot 14 \times 10^{-21} \times 200 \times 10^3$$

or $$N_i = 8 \cdot 28 \times 10^{-16}$$

Now $F = N_o/N_o'$ where N_o is the output noise of the amplifier and N_o' is the output noise of an ideal amplifier. Hence

$$F = \frac{1200 + 700}{1200} = 1 \cdot 58$$

Also $$F = \frac{S_i/N_i}{S_o/N_o} = S_i/N_i \qquad (\text{since } S_o/N_o = 1)$$

Hence $$e_s^2/300 = 1 \cdot 58 \times 8 \cdot 28 \times 10^{-16}$$

or $$e_s^2 = 0 \cdot 392 \times 10^{-12}$$

and $$e_s = 0 \cdot 63 \, \mu\text{V}$$

Example 5.2

Define the terms noise factor and signal-to-noise ratio and explain a relationship between them in respect of a two-port network.

Figure 5.3 shows a two-port network with its noise properties represented by two equivalent input generators, one of mean-square voltage $\overline{v^2}$ and the other of mean-square current $\overline{i^2}$. The network is driven from a source having an internal resistance R_g. Assuming the generators to be uncorrelated, deduce an expression for the noise factor of the network with this source and show that it is a minimum when $R_g^2 = \overline{v^2}/\overline{i^2}$. Give expressions for $\overline{v^2}$ and $\overline{i^2}$ if the network consists of (a) a single shunt-connected resistor R or (b) a single series-connected resistor R. (U.L.)

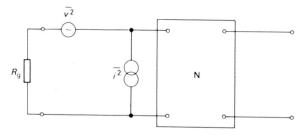

Fig. 5.3

Solution
The answer to the first part of the question is given in Section 5.1.

Problem

For the equivalent circuit shown in Fig. 5.4, if v_n is the total noise voltage at the input then

$$\overline{v_n^2} = \overline{v_g^2} + \overline{v^2} + \overline{i^2}\, R_g^2$$

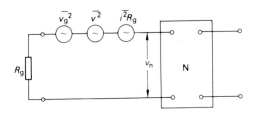

Fig. 5.4

Without the network, the only noise source is R_g which produces a mean-square noise voltage $\overline{v_g^2} = 4kTBR_g$. Hence

$$F = \frac{\overline{v_n^2}}{\overline{v_g^2}} = 1 + \frac{\overline{v^2}}{4kTBR_g} + \frac{\overline{i^2}\,R_g}{4kTB}$$

For a minimum, $dF/dR_g = 0$ and we obtain

$$0 = -\frac{\overline{v^2}}{4kTBR_g^2} + \frac{\overline{i^2}}{4kTB}$$

or

$$R_g^2 = \overline{v^2}/\overline{i^2}$$

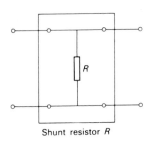

Shunt resistor R

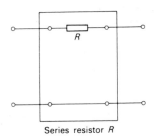

Series resistor R

Fig. 5.5

(a)   Shunt resistor R

Here

$$\overline{v^2} = 0$$

Hence

$$\overline{i^2} = 4kTB/R$$

(b)   Series resistor R

Here

$$\overline{i^2} = 0$$

Hence

$$\overline{v^2} = 4kTBR$$

5.2   Typical noise factors

The most common circuits used are the grounded-cathode and grounded-grid circuits for valves, the corresponding common-emitter and common-base circuits for transistors, and circuits for the JFET and IGFET. Further details are given in Appendix K and the results are summarised below.

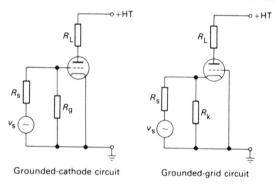

Fig. 5.6

Grounded-cathode circuit

$$F = 1 + \frac{R_s}{R_g} + \frac{R_{eq}}{R_s}\left(1 + \frac{R_s}{R_g}\right)^2$$

where R_{eq} is the equivalent shot noise resistance

or

$$F = 1 + R_{eq}/R_s$$

if $R_g \gg R_s$.

Grounded-grid circuit

$$F = 1 + \frac{R_s}{R_g} + \left(\frac{\mu}{\mu+1}\right)^2\frac{R_{eq}}{R_s}\left(1 + \frac{R_s}{R_g}\right)^2$$

where μ is the amplification factor of the valve

or

$$F = 1 + R_{eq}/R_s$$

if $R_g \gg R_s$ and $\mu \gg 1$.

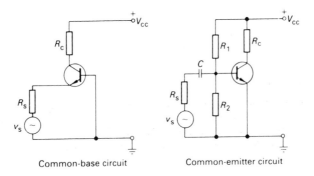

Fig. 5.7

Common-base circuit

$$F = 1 + \frac{r_e}{2R_s} + \frac{r_b}{R_s} + \frac{(1-\alpha_0)(R_s + r_e + r_b)^2 \left[1 + \left(\dfrac{f/f_{ab}}{\sqrt{1-\alpha_0}}\right)^2\right]}{2\alpha_0 r_e R_s}$$

where R_s is the source resistance, r_e and r_b are the emitter and base resistances respectively, α_0 is the d.c. current-gain factor, and f_{ab} is the cut-off frequency of the common-base circuit.

Common-emitter circuit

$$F = 1 + \frac{r_e}{2R_s} + \frac{r_b}{R_s} + \frac{(1-\alpha_0)(R_s + r_e + r_b)^2 \left[1 + \left(\dfrac{f/f_{ae}}{\sqrt{1-\alpha_0}}\right)^2\right]}{2\alpha_0 r_e R_s}$$

where R_s is the source resistance, r_e and r_b are the emitter and base resistances respectively, α_0 is the d.c. current-gain factor, and f_{ae} is the cut-off frequency of the common-emitter circuit.

JFET circuit

$$F = 1 + \frac{g_i}{g_s} + \frac{R_n(g_i + g_s)^2}{g_s}$$

where g_i is the conductance of a noise current generator, g_s is the source conductance, and R_n is an equivalent noise resistance.

IGFET (MOSFET) circuit

$$F = 1 + R_n g_{in}/g_m$$

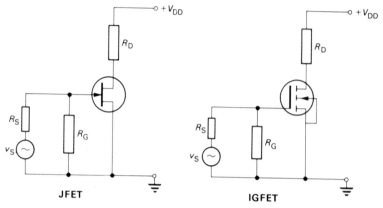

Fig. 5.8

where R_n is an equivalent noise resistance, g_{in} is the input conductance, and g_m is the forward transconductance.

5.3   Cascaded networks

When two or more active networks such as amplifiers are connected in cascade to give greater amplification, the overall noise factor F of the arrangement is important.

Consider two networks in cascade with noise factors of F_1 and F_2 and power gains of G_1 and G_2 respectively. Let F be the overall noise factor of the combined networks with a relevant bandwidth B. If a noise signal with power kTB is fed into the first network we have

$$\text{output noise of first network} = F_1 G_1 kTB$$

$$\text{output noise of ideal network} = G_1 kTB$$

$$\text{noise generated in first network} = F_1 G_1 kTB - G_1 kTB$$

$$= (F_1 - 1)kTBG_1$$

The noise generated in the first network therefore appears to be due to a hypothetical input noise component which is equal to $[(F_1 - 1)kTBG_1/G_1]$ $= (F_1 - 1)kTB$ and is shown dotted in Fig. 5.9. Similarly, the equivalent noise generator associated with the second network is $(F_2 - 1)kTB$. Hence, the first two networks can be regarded as 'ideal' for the purposes of this analysis if the noise contributions from the *two* input signals are taken into account and the total noise output from the second network $= kTBG_1 G_2 + (F_1 - 1)kTBG_1 G_2$ $+ (F_2 - 1)kTBG_2 = F_1 kTBG_1 G_2 + (F_2 - 1)kTBG_2$.

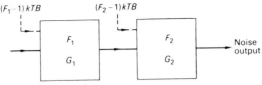

Fig. 5.9

Now the noise power output from two *ideal* networks in cascade equals $kTBG_1G_2$. Hence, the noise factor F of the combined networks is given by

$$F = \frac{\text{noise output of two given networks}}{\text{noise output of two ideal networks}}$$

$$= \frac{F_1 kTBG_1G_2 + (F_2 - 1)kTBG_2}{kTBG_1G_2}$$

or
$$F = F_1 + (F_2 - 1)/G_1$$

Comments

1. For three networks in cascade we obtain similarly

$$F = F_1 + \frac{(F_2 - 1)}{G_1} + \frac{(F_3 - 1)}{G_1 G_2}$$

2. The noise factor of an attenuator with a loss L can be shown to be also L, i.e. $F = L$ (numerically).

5.4 Noise factor measurement[25]

Two methods for measuring the noise factor of an amplifier or receiver are the small signal method and the noise diode method. The latter is more convenient to use as it does not involve a knowledge of the bandwidth of the system, which is generally difficult to know accurately.

The noise diode method uses as a noise source a temperature-limited diode which generates sufficient noise over a broad band of frequencies up to about 5 MHz. The noise is essentially shot noise and it may be varied by varying the diode current. The noise diode is shown in Fig. 5.10 and a typical arrangement is given in Fig. 5.11.

The noise diode filament is heated by a variable voltage source to operate in the temperature-limited region. In the anode circuit there is a large RF choke which has a high impedance at the mid-band frequency of interest and its impedance can be neglected as can the high diode impedance r_d. Hence, the

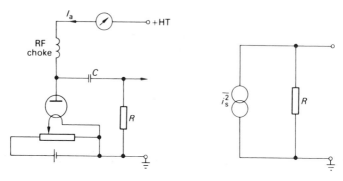

Fig. 5.10

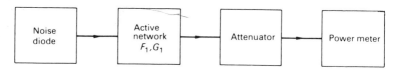

Fig. 5.11

equivalent circuit in Fig. 5.10 is a shot noise current generator delivering a shot noise mean-square current $\overline{i_s^2}$.

Let the input resistance to the active network which is used be R_g. The measurement consists in noting the output power P_o with the diode current zero initially and with zero attenuation in circuit. The diode current is then adjusted to a value I_a which gives the same output reading with 3 dB attenuation in circuit, i.e. the output power has been doubled.

To obtain a suitable meter reading for P_o, the network gain is set at the *same* convenient value for both readings. The use of 3 dB attenuation maintains the reading at P_o in both cases and avoids any scale non-linearity. Hence, we have

$$\text{noise power available from resistor } R_g = \frac{\overline{e_t^2}}{4R_g} = kTB$$

$$\text{noise power output of diode} = \frac{\overline{e_s^2}}{4R_g} = \frac{eI_aR_gB}{2}$$

Hence, when $I_a = 0$ we have

$$P_o = kTBGF \qquad \text{(since } F = N_o/kTBG)$$

When the diode current is set at the value I_a we have

$$P_o + P_o = kTBGF + eI_a R_g BG/2$$

or

$$2kTBGF = kTBGF + eI_a R_g BG/2$$

Hence

$$kTBGF = eI_a R_g BG/2$$

with

$$F = eI_a R_g/2kT$$

Typically, $e/2kT \simeq 20$, hence

$$F \simeq 20 I_a R_g$$

and it can easily be evaluated since the values of I_a and R_g are known directly in the measurement.

Comment
At higher frequencies in the microwave region, the diode noise power is insufficient for measuring large noise factors and so a gas discharge tube is used. It is placed at a small angle across the waveguide to produce an impedance match. At present, solid state noise sources are also available. (See Section 5.9.)

5.5 Noise temperature

The thermal noise power P_n available from a resistor in a bandwidth B is kTB where T is the absolute temperature of the resistor. Hence, an alternative concept associated with noise power is the *effective noise temperature* T_e which is given by

$$T_e = P_n/kB$$

for a resistor at temperature T K.

The idea of effective noise temperature can be extended to other noisy sources which are not necessarily associated with a *physical* temperature as is the case with a resistor. For example, a non-thermal device, e.g. an antenna which picks up noise power due to the random radiation it receives from various directions, may also be associated with an effective noise temperature T_a. If the noise power received by the antenna is P_n in a bandwidth B then

$$T_a = P_n/kB$$

The value of T_a depends on the direction in which the antenna points and its radiation pattern. Different parts of the sky are associated with sources of random radiation usually called galactic noise, solar noise, etc. Hence, the sky *effectively* has a 'noise temperature' and it varies with frequency as shown in Fig. 5.12.

The concept of effective noise temperature may also be applied to an active network, such as an amplifier. It is found to be more useful and meaningful for low-noise amplifiers, such as masers or parametric amplifiers. In the case of

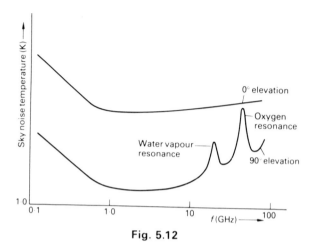

Fig. 5.12

these low-noise amplifiers, their noise factors may not be very different from unity and, for comparative purposes, it is usual to compare their effective noise temperatures. If T_e is the effective noise temperature of a network with a noise factor F, the relationship between them is obtained as follows.

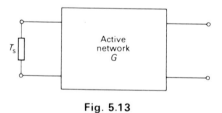

Fig. 5.13

Consider a thermal noise source at temperature T_s connected to the input of the network of Fig. 5.13. The output noise power is $FGkT_sB$ and the noise power generated by the network is $FGkT_sB - GkT_sB = (F - 1)GkT_sB$. If the network is assumed to behave as a noisy resistor with an effective input noise temperature T_e, then the output noise power it produces is GkT_eB and we have

$$GkT_eB = (F - 1)GkT_sB$$

or

$$T_e = (F - 1)T_s$$

and

$$F = 1 + T_e/T_s$$

The source temperature here is also the *input* temperature of the active network and the standard value taken is 290 K which yields

$$T_e = (F - 1)290 \text{ K}$$

or

$$F = 1 + T_e/290$$

If the signal source temperature T_s is different to the standard input temperature $T_0 = 290$ K of the network, the noise factor F is given by

$$F = 1 + (F_{290} - 1)T_0/T_s$$

where F_{290} is the normally defined noise factor of the network.

Comments
1. For active networks in cascade with effective noise temperatures of T_{e_1}, T_{e_2}, etc. and power gains of G_1, G_2, etc. respectively, the overall noise temperature T_e is given by

$$T_e = T_{e_1} + \frac{T_{e_2}}{G_1} + \frac{T_{e_3}}{G_1 G_2} + \ldots$$

   with $T_{e_1} = (F_1 - 1)290$ K, $T_{e_2} = (F_2 - 1)290$ K, etc. and F_1, F_2, etc. are defined at 290 K.
2. For a lossy line, attenuator, or radome, with a loss factor L, we obtain $L = F$ with $T_e = (L - 1)290$ K.
3. A small difference in noise factors between 1·1 and 1·2 gives a larger difference in noise temperatures between 29 K and 58 K which is easier to handle in low-noise systems.

Example 5.3
Three matched amplifiers are available to amplify a low-level signal. They have the following characteristics.

Amplifier	Power gain	Noise factor
A	6 dB	1·7
B	12 dB	2·0
C	20 dB	4·0

The amplifiers are to be connected in cascade. Calculate from first principles the lowest overall noise factor obtainable, noting the order in which the amplifiers must be connected.

Compare and contrast noise factor and noise temperature, stating the particular advantages and applications of each. (C.E.I.)

Solution
To obtain the lowest overall noise factor, the first amplifier should have the lowest noise factor, i.e. amplifier A should be the first one. For the second amplifier, there are two possibilities, B or C. Hence, the arrangement is either (a) A, B, C or (b) A, C, B.

(a)   For A, B, C

$$F = F_A + \frac{(F_B - 1)}{G_A} + \frac{(F_C - 1)}{G_A G_B}$$

$$= 1.7 + \frac{(2 - 1)}{4} + \frac{(4 - 1)}{4 \times 16}$$

$$= 1.7 + 0.25 + 0.047$$

or $F = 1.997$

(b)   For A, C, B

$$F = F_A + \frac{(F_C - 1)}{G_A} + \frac{(F_B - 1)}{G_A G_C}$$

$$= 1.7 + \frac{(4 - 1)}{4} + \frac{(2 - 1)}{4 \times 100}$$

$$= 1.7 + 0.75 + 0.025$$

or $F = 2.475$

Hence, the optimum arrangement is amplifier A first, then amplifier B, and, finally, amplifier C giving an overall value of $F = 1.997$.

Noise factor and noise temperature comparison
1.   Both give a measure of the 'noisiness' of the network; the larger they are, the greater is the 'noisiness'.
2.   They are inter-related by the expression

$$T_e = (F - 1)290 \quad \text{or} \quad F = 1 + T_e/290$$

3.   Both are associated with noise power and give its measure.
4.   Both can be applied to cascaded networks to give the total effect.

Noise factor and noise temperature contrast
1. F is a pure number while T_e is a physical temperature in some cases.
2. $F_{min} = 1$ while $T_{min} = 0$K.
3.   It is convenient to use large values of F but, for low values of F, T_e is larger in magnitude and range. Hence, T_e is more useful in low-noise systems.

The main application of F is in amplifier and receiver noise performance, especially in cascaded arrangements at normal room temperatures. The main application of T_e is in low-noise systems,[26] such as maser or parametric preamplifiers. It is also applied to certain antennas, such as parabolic reflectors, and for expressing sky noise temperature.

5.6   System noise temperature

An important parameter used in low-noise applications, such as a satellite communications ground station, is the system noise temperature T_s. For the case of an antenna connected directly to a receiver, T_s is defined by

$$T_s = T_a + T_r$$

where T_a is the antenna noise temperature associated with its radiation resistance in thermal equilibrium with the environment and T_r is the effective *input* noise temperature of the receiver due to its internal noise sources.

However, for convenience, a transmission line often connects the antenna to a receiver and gives rise to thermal losses, while a preamplifier may be used ahead of the receiver to improve system performance. In this case, it is convenient to refer noise temperatures of the 'receiver chain' to a point just behind the antenna, as shown in Fig. 5.14. The system noise temperature T_s is then given by

$$T_s = T_a + (L-1)T_0 + LT_p + LT_r/G$$

where L is the loss factor of the transmission line, T_0 is the ambient temperature, T_p is the noise temperature of the preamplifier with power gain G, and T_r is the effective noise temperature of the receiver.

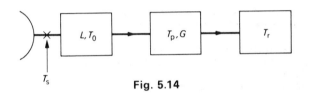

Fig. 5.14

Example 5.4
An antenna with a noise temperature of 57 K is connected by means of a cable to a preamplifier and receiver. The cable loss is 1 dB and the preamplifier has a 20 dB power gain and a noise temperature of 90 K. If the effective temperature of the receiver is 290 K, what is the system noise temperature? Assume an ambient temperature of 290 K.

Solution
We have

$$T_s = T_a + (L-1)T_0 + LT_p + LT_r/G$$

with $$L = 1\,\text{dB} = 1{\cdot}26$$

and $$G = 20\,\text{dB} = 100$$

Hence

$$T_s = 57 + (1{\cdot}26 - 1{\cdot}0)290 + (1{\cdot}26 \times 290) + (1{\cdot}26 \times 290)/100$$

$$= 57 + 75{\cdot}4 + 113{\cdot}4 + 3{\cdot}65$$

or $$T_s = 249{\cdot}5\,\text{K}$$

Comments
1. The large contribution to T_s is due to the loss factor L. Hence, it is very important to reduce line losses to a minimum.
2. By using a preamplifier with high gain, the large noise temperature of the receiver degrades performance by only $3.65\,K$.

5.7   Low-noise amplifiers[27, 28]

In radar systems and satellite communications, the power of the received signal is very low and it is necessary to provide further amplification at a low-noise level. Three well-known devices used are the maser, parametric amplifier, and FET amplifier.

Maser
This is a device capable of amplifying weak microwave signals. The word maser is an abbreviation for 'microwave amplification by the stimulated emission of radiation' and masers are usually of the gaseous or solid-state forms. A well-known solid-state maser is the ruby maser which is used either as a cavity maser or as a travelling-wave maser.

The operation of the maser is based on quantum-mechanical principles. Ruby is the aluminium oxide Al_2O_3 which is embedded with some Cr ions and, when placed in an external magnetic field, splits up into three or four energy levels, as shown in Fig. 5.15. At room temperature, the electrons are mainly in the lower energy levels according to the Boltzmann equation

$$\frac{N_1}{N_2} = \frac{e^{-W_1/kT}}{e^{-W_2/kT}}$$

such that $N_1 > N_2$ and $W_1 < W_2$.

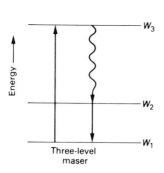

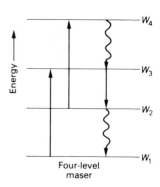

Fig. 5.15

To produce amplification, electrons must be raised to energy level W_3 (three-level maser) by applying external RF energy at the 'pump' frequency $f_p = (W_3 - W_1)/h$. This is known as *population inversion* and the excited electrons return spontaneously to energy level W_2 due to spin-lattice relaxation and phonon emission. On application of a weak input signal at frequency f_s, a large number of coherent electron transitions are *stimulated* from energy level W_2 to the ground state W_1.

Hence, the emission of energy occurs at the frequency $f_s = (W_2 - W_1)/h$ which produces signal amplification. However, to prevent spontaneous incoherent transitions to the lower levels which give rise to noise, the ruby crystal must be cooled in liquid helium or nitrogen.

In the cavity maser, the ruby crystal is placed at one end of a resonant microwave cavity while signal and pump power is coupled in at the other. This structure has a narrow bandwidth and is overcome by using the travelling-wave maser which employs a slow-wave structure. Interaction between the ruby crystal and RF field can take place over a greater length and it achieves high gain with a wide bandwidth.

Masers are still used in such applications as radio astronomy or deep space communications but have been superseded in other areas by the parametric amplifier or paramp. Typically, a travelling-wave maser operating around 1–2 GHz has a gain of 20 dB, a bandwidth of 25 MHz, and a noise temperature of 20 K. The pump frequency used is of the order of 48 GHz and has a power level of about 150 mW. The travelling-wave maser is illustrated in Fig. 5.16.

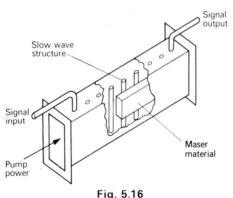

Fig. 5.16

Parametric amplifier

In this device, amplification is produced by varying the capacitance parameter of a varactor diode which forms part of a tuned circuit. Hence, the name parametric amplifier or paramp. The capacitance variation is produced by the energy from a pump source. By varying the distance between the capacitor 'plates' at suitable moments during the input signal, the voltage across the varactor diode increases and so produces amplification of the input signal, as shown in Fig. 5.17.

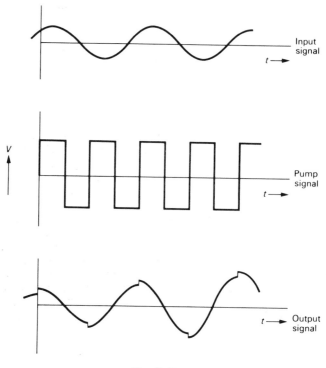

Fig. 5.17

A typical paramp using a three-port circulator is shown in Fig. 5.18. It is coupled to an input signal circuit, an idler circuit, and the pump circuit across which is the varactor diode. The pump frequency f_p, the idler frequency f_i, and the signal frequency f_s are related by $f_i = f_p - f_s$ and, if $f_p = 2f_s$, $f_i = f_s$ and the device is called a *degenerate* paramp. It requires only two circuits as the idler and input circuits are identical.

More generally, to produce sufficient gain, a non-degenerate paramp is used with an idler circuit, in which case $f_p \neq 2f_s$ and amplification occurs at the idler frequency $f_i = f_p - f_s$. The device can then be operated also as an up-converter or down-converter.

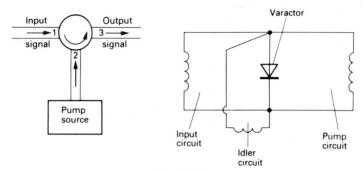

Fig. 5.18

The amplification, which occurs at the idler frequency if the pump frequency is higher than the signal frequency, is given by the Manley–Rowe relations.[29] It is due to the presence of a negative resistance at the varactor diode and stability is possible if the negative resistance is less than the signal source resistance. However, the single varactor paramp has a narrow bandwidth of around 10 % of the centre frequency and, to increase the bandwidth to 50 % of the centre frequency, a travelling-wave structure is used with several varactor diodes.

The performance of the paramp depends on the operating frequency, pump frequency, and varactor characteristic. The noise factor can be quite low and is reduced by cooling with liquid nitrogen (77 K) or liquid helium (4·2 K). Typically, a paramp operating at 5 GHz achieves a gain of 20 dB with a bandwidth of 10 % using a Gunn diode pump source and a GaAs varactor. At room temperature, the noise temperature is about 70 K; it can be as low as 20 K when cooled with liquid helium. Paramps are used in radar systems and for satellite communications at ground stations.

FET amplifier[30]
Field-effect transistors made from GaAs have achieved significant results at microwave frequencies. Microwave GaAs field-effect transistors and, in particular, low-noise FET amplifiers have been developed during recent years and are now commercially available.

Their high-frequency performance is due to GaAs which has a higher electron mobility and greater peak electron velocity than silicon and these lead to a faster transit time and lower dissipation. GaAs FETs are now able to operate at microwave frequencies beyond 40 GHz with high gain and low noise.

In a GaAs FET, charge carriers flow from source to drain through a thin epitaxial layer of n-type GaAs on the surface of semi-insulating GaAs bulk material. The current flow is controlled by either a Schottky barrier gate or an insulated gate electrode. The Schottky barrier FET has superior microwave performance and is fabricated by the planar process. The thin conducting GaAs epitaxial layer is deposited on a semi-insulating GaAs substrate and the Schottky barrier gate is formed by the deposition of a metal on the epitaxial layer. The construction of a typical device is shown in Fig. 5.19.

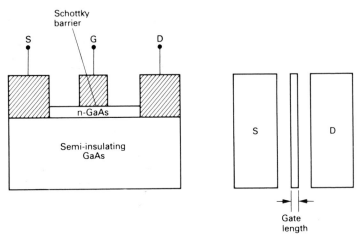

Fig. 5.19

At present, the most important application of GaAs FETs is in low-noise amplifiers. GaAs FETs designed for low-noise applications have Schottky barrier gates with typical gate length of 0·5 μm. Noise figures range from 1 dB at 4 GHz to about 2 dB at 12 GHz with a power gain of about 10–12 dB.

A microwave low-noise amplifier may employ one or two GaAs FET stages followed by a bipolar amplifier. Alternatively, the first stage can be an ultra low-noise parametric amplifier followed by a low-noise FET stage. An overall noise figure of 3 dB can be achieved over a 1 MHz band in the frequency range 7·25–7·75 GHz.

FET amplifiers have other advantages, such as long shelf and working lives, small size, and low power dissipation. They are finding increasing use in the front end of various kinds of microwave receivers for both radar and satellite communications. In earth station applications, they are a very reliable and cost-effective way of implementing a low-cost satellite ground station with an excellent figure of merit G/T.

Example 5.5
Sketch and discuss the variation of the sky noise temperature as a function of frequency.
 A receiver has a system noise factor of 10 dB and it is proposed to improve its sensitivity by adding a preamplifier of 3 dB noise factor and 10 dB power gain.

To what extent will the receiver performance improve if the effective noise temperature of the receiver aerial is (a) 290 K or (b) 58 K? (Quoted noise factors are referred to 290 K.)   (C.E.I.)

Solution
The sky noise temperature shown in Fig. 5.12 is due to various galactic sources and the sun. Noise power from the sun decreases inversely as frequency at the lower frequencies up to about 1 GHz. In the microwave region between 1 GHz and 10 GHz, the noise power received by an antenna is less than at other frequencies. This is known as the low-noise 'window' which is useful for space communications. Above 10 GHz, the sky noise temperature increases, with resonance peaks at about 22 GHz, due to water vapour, and at about 60 GHz, due to O_2.

It is also observed that the noise power received by the antenna is a minimum when it is pointing towards the zenith and is a maximum when it is pointing towards the horizon. In the latter case, its side lobes pick up random radiation from the earth which behaves as a 'hot body'. This is illustrated by the two curves shown in Fig. 5.12.

Problem
Let the *system* noise temperature of the aerial and amplifier be T_s. For each value of aerial noise temperature, the effective noise temperature of the system with and without a preamplifier will be compared to show the improvement in performance.

(a)   For 290 K
without preamplifier $T_s = 290 + (10 - 1)290 = 2900$ K

with preamplifier $T_s = 290 + (2 - 1)290 + \dfrac{(10 - 1)290}{10} = 841$ K

$$\text{improvement factor} = \frac{2900}{841} = 3{\cdot}45 = 5{\cdot}4 \text{ dB}$$

(b)   For 58 K
without preamplifier $T_s = 58 + (10 - 1)290 = 2668$ K

with preamplifier $T_s = 58 + (2 - 1)290 + \dfrac{(10 - 1)290}{10} = 609$ K

$$\text{improvement factor} = \frac{2668}{609} = 4{\cdot}4 = 6{\cdot}4 \text{ dB}$$

Hence, there is a further improvement in the latter case of 1 dB when the aerial noise temperature is 58 K.

Example 5.6
A receiver gives satisfactory results when the signal-to-noise power ratio at the output of the linear stages is better than 40 dB. The noise figure of the input stage of the receiver is 10 dB and it has a loss of 8 dB. The following stage has a noise figure of 3 dB and has high gain. At a particular site, the input noise temperature is 7 °C and the input signal-to-noise power ratio is 1×10^5. Is the receiver in its present form satisfactory under these

conditions or is a preamplifier necessary? If the latter, what should be the main points in its specification?

Assume that the bandwidth of all the linear stages is the same and noise figures are defined with reference to 290 K. (C.E.I.)

Solution
The network arrangement of the receiver is shown in Fig. 5.20.

The overall noise factor of the receiver when referred to $T = 290$ K is

$$F_{290} = F_1 + \frac{F_2 - 1}{G_1}$$

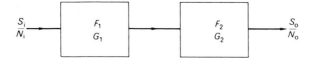

Fig. 5.20

with $F_1 = 10$, $F_2 = 2$, and $G_1 = 1/L = 1/6 \cdot 31$ where L is the loss. Hence

$$F_{290} = 10 + \frac{(2 - 1)}{(1/6 \cdot 31)} = 16 \cdot 31$$

When referred to the *input* temperature of $7\,^{\circ}\mathrm{C}$ ($T = 280$ K) we have

$$F_{280} = 1 + (F_{290} - 1)\frac{290}{280} = 1 + 15 \cdot 31 \times \frac{290}{280}$$

or $$F_{280} = 16 \cdot 8$$

Also $$F_{280} = \frac{S_i/N_i}{S_o/N_o}$$

or $$S_o/N_o = \frac{10^5}{16 \cdot 8} < 10^4 \quad \text{or} \quad 40 \text{ dB}$$

Hence, the receiver is unsatisfactory in its present form and requires a preamplifier. The specifications of the latter must be such that the overall noise factor of the arrangement is < 10. If F is the noise factor of the preamplifier and G is its gain we have

$$F_{290} = F + \frac{9}{G} + \frac{6 \cdot 31}{G}$$

By inspection, when $F = 8$ and $G = 10$ we obtain $F_{290} = 9 \cdot 53$ and $F_{280} = 9 \cdot 84$ or $F_{280} < 10$. Hence, when $F \leqslant 8$ and $G \geqslant 10$ we obtain $S_o/N_o > 10^4$ or 40 dB.

5.8 Noise temperature measurement[31]

The effective noise temperature T_a of an antenna is usually measured using the 'Y-factor method'. The principle of the method is to compare the noise power received by the antenna to the noise power generated by a standard noise source and, from the ratio of these noise powers, T_a can be determined. The circuit arrangement used is shown in Fig. 5.21.

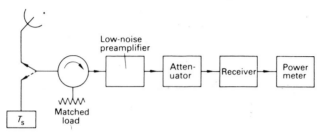

Fig. 5.21

With the standard liquid-cooled load connected to the input, the output noise power N_s is noted on the meter. The antenna is now connected to the input and the output noise power N_a is noted on the meter. If the ratio N_s/N_a is denoted by Y and the receiver noise temperature is T_R we have

$$Y = \frac{N_s}{N_a} = \frac{k(T_s+T_R)B}{k(T_a+T_R)B}$$

where B is the relevant system bandwidth. Hence

$$Y = \frac{T_s+T_R}{T_a+T_R}$$

or

$$YT_a + YT_R = T_s + T_R$$

and

$$T_a = \frac{T_s+T_R(1-Y)}{Y}$$

To measure T_R, the output noise power N_h of a 'hot load', i.e. a load at room temperature, is compared with the noise power output N_s of the standard cooled load. If the ratio of these noise powers is Y' then we have

$$Y' = \frac{N_h}{N_s} = \frac{k(T_h+T_R)B}{k(T_s+T_R)B}$$

with

$$Y'T_s + Y'T_R = T_h + T_R$$

or

$$Y'T_R - T_R = T_h - Y'T_s$$

Hence
$$T_R = \frac{T_h - Y'T_s}{(Y' - 1)}$$

which can be determined since Y', T_h, and T_s are known by direct measurement. The value of T_a can then be determined from the previous expression using the known values of T_s, T_R, and Y.

5.9   Excess noise ratio (ENR)[32]

Microwave noise generators are usually of the solid-state or gas discharge types which use room temperature as the *standard* reference by simply de-energising the noise source and assuming room temperature as the standard value T_0. In this case, the effective receiver temperature T_R is related to the noise factor F by

$$T_R = \frac{T_h - Y'T_0}{(Y' - 1)} = (F - 1)T_0$$

with
$$F - 1 = \frac{(T_h/T_0) - Y'}{(Y' - 1)}$$

or
$$F = \frac{(T_h/T_0) - 1}{(Y' - 1)}$$

where T_h is the hot-load temperature of the energised source.

The quantity in the numerator is a measure of the power output of the noise source and is called the *excess noise ratio* (ENR) which is given by

$$\text{ENR} = 10 \log (T_h/T_0 - 1)\,\text{dB}$$

and
$$F = \text{ENR} - 10 \log(Y' - 1)\,\text{dB}$$

Solid-state noise sources are semiconductor p–n diodes operating in the avalanche region. The randomness of the avalanche multiplication process produces fluctuations in the avalanche current which generate random noise over a wide frequency range. Typically, the diodes operate at voltages of around 20 to 30 volts and are driven from a constant current source from between 5 to 20 mA.

The choice between solid-state and gas discharge noise sources is based on frequency coverage. For laboratory work below 18 GHz requiring operation at many frequencies, solid-state units offer economic advantages, in addition to small size, low mass, and low power consumption. Usually, solid-state noise sources are calibrated at several frequencies and, typically, may have an ENR of around 15 dB in the frequency range 1–12·4 GHz or, in certain cases, an ENR as high as 40 dB. They can be used to determine the noise figures of amplifiers, mixers, or receivers and also to check the performance of radar and communication systems.

6
Systems

Signals transmitted by any communication system will always be received with a certain amount of noise present, as the latter cannot be entirely eliminated from the system. Hence, in evaluating systems performance, or when comparing different communication systems, the most generally used criterion is the output *signal-to-noise ratio* of the system as mentioned in an earlier chapter.

In many cases, the most usual definition of signal-to-noise ratio is that of *average* signal power to *average* noise power present in the system though, sometimes, other definitions may be used. The signal-to-noise ratio will often be written as S/N in this chapter and it is useful to compare its value at the input and output ends of the receiving system which will be designated as (S_i/N_i) and (S_o/N_o) respectively.

Most communication systems employ either analogue or pulse modulation and the evaluation of the S/N ratio leads to different values for various systems. The performance of any system may be judged in terms of these values and it leads to the conclusion that, for best results, the S/N ratio of the system must be as large as possible. Poor S/N ratios mainly signify that the presence of noise is serious and greatly reduces the usefulness of the system, while a good S/N ratio would indicate that the effect of noise in the system is quite small and is not important.

An analysis of typical S/N ratios for analogue systems such as AM, DSBSC, SSB, and FM will first be undertaken and a comparison made of their relative merits. Subsequently, pulse systems such as PAM, PDM, and PPM will be considered and a similar comparison will be made. In the case of PCM, it is more practical to consider the disturbing effect of errors due to quantisation noise rather than the other types of noise present in the system and this will be undertaken using probability theory. Furthermore, some idea of communication rate will also be considered as a useful means of signifying merit in a particular system. Finally, as mentioned earlier, it is sometimes more useful to consider the ratios E/N_0 and C/N_0. These ratios will be considered when dealing with digital and satellite communication systems at the end of this chapter.

6.1 Analogue systems[33, 34]

The most familiar systems are amplitude modulation (AM), double-sideband suppressed carrier (DSBSC), single-sideband suppressed carrier (SSBSC), and frequency modulation (FM).

AM system

The carrier and both sidebands are present in this system and, for a *large* signal-to-noise ratio S_i/N_i at the input to the receiver, it is useful to consider the corresponding ratio S_o/N_o obtained at the output of the detector or demodulator. In this case, the most common detector used is the envelope detector for S_i/N_i ratios greater than about 10 dB. The expression for an AM signal at the detector input is given by

$$v_i(t) = V_c(1 + m \sin \omega_m t)\sin \omega_c t$$

where V_c is the peak carrier voltage of frequency $\omega_c/2\pi$ and m is the depth of modulation at the modulating frequency $\omega_m/2\pi$. Hence, the average input *sideband* power into the envelope detector is given by

$$S_i = \frac{m^2}{2} \times \frac{V_c^2}{2} = \frac{m^2 V_c^2}{4}$$

For an IF bandwidth of $2B$, where B is equal to the highest modulating frequency, the output voltage from the envelope detector (assumed to be ideal) is

$$v_o(t) = m V_c \sin \omega_m t$$

and the average signal power associated with this output is

$$S_o = m^2 V_c^2/2$$

Assuming the input noise has a uniform spectral distribution such as 'white noise', noise frequency components within a bandwidth δf can be treated as a *single* frequency component at the centre of the band δf with the same total noise power. This is illustrated in Fig. 6.1.

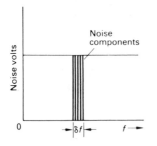
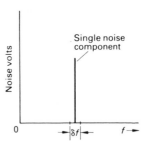

Fig. 6.1

If $S(f)$ is the noise power spectral density, then the power in a bandwidth δf associated with the single 'impulse' is $S(f)\,\delta f = N_0\,\delta f$ where N_0 is the power spectral density in watts/Hz. If the corresponding single noise component is $v_n(t)$ we have

$$v_n(t) = V_n \sin 2\pi (f_{IF} + f_n)t$$

where f_{IF} is the IF frequency and f_n is the frequency of the particular noise frequency component with a peak value V_n and $N_0\,\delta f = V_n^2/2$.

When $\delta f \to 0$, all the noise components cover the IF bandwidth $2B$ continuously and the total noise power N_i is

$$N_i = \int_{-B}^{+B} N_0\,\delta f = 2N_0 B$$

and

$$\frac{S_i}{N_i} = \frac{m^2 V_c^2}{4} \bigg/ 2N_0 B = \frac{m^2 V_c^2}{8N_0 B} = \frac{m^2 P_c}{4N_0 B}$$

where P_c is the *average* carrier power.

To obtain the output noise power of the detector, each noise component $v_n(t)$ within a bandwidth δf will beat with the carrier and the resultant voltage is given approximately by

$$v_n(t) = V_c [1 + (V_n/V_c)\cos \omega_n t]\sin \omega_c t$$

if $V_c \gg V_n$. When this is applied to the detector, the *output* noise voltage $n_o(t) = V_n \cos \omega_n t$. Hence, the average noise power δN_o in a bandwidth δf is

$$\delta N_o = V_n^2/2 = N_0\,\delta f \qquad \text{(as obtained earlier)}$$

and for the whole IF bandwidth this becomes

$$N_o = \int_{-B}^{+B} N_0\,\mathrm{d}f = 2N_0 B$$

Hence

$$\frac{S_o}{N_o} = \frac{m^2 V_c^2}{2} \bigg/ 2N_0 B = \frac{m^2 V_c^2}{4N_0 B} = \frac{m^2 P_c}{2N_0 B}$$

which is equal to twice S_i/N_i or

$$(S_o/N_o)_{AM} = 2(S_i/N_i)_{AM}$$

and amounts to a 3 dB improvement at the detector. The 3 dB improvement is due to arithmetic addition of power in the sidebands and quadratic addition of the independent noise in each sideband.

DSBSC system

The only difference between an AM system and a DSBSC system is the carrier power which is present in the former. Hence, for *equal* average power in the *sidebands*, the S_o/N_o ratios of the two systems must be the same, which yields

$$(S_o/N_o)_{DSBSC} = (S_o/N_o)_{AM}$$

and

$$(S_o/N_o)_{DSBSC} = 2(S_i/N_i)_{DSBSC}$$

if a coherent detector is used.

For equal *total* powers, however, the carrier power in an AM system will help to increase the sideband powers in the DSBSC system. Hence, we have

$$(S_o/N_o)_{DSBSC} = 3(S_o/N_o)_{AM}$$

for 100 % modulation, since the total AM power is equal to three times the sideband power. However, in practice, the main limitation at the transmitter is on *peak* power rather than average power. Hence, on the basis of peak power, it can be shown that

$$(S_o/N_o)_{DSBSC} = 4(S_o/N_o)_{AM}$$

which amounts to a 6 dB improvement.

SSB system

The expression for an SSB signal is $v(t) = V_m \cos(\omega_c \pm \omega_m)t$. On passing this signal through the product detector, its RF frequency components (signal and noise) are changed to the lower audio frequencies. Hence, there is only a frequency change of the signal and noise components between input and output and so we obtain

$$(S_o/N_o)_{SSB} = (S_i/N_i)_{SSB}$$

In comparison with a DSBSC system, since the noise power input N_i is reduced by half, due to only one set of sidebands being present, we must have

$$(S_i/N_i)_{SSB} = 2(S_i/N_i)_{DSBSC}$$

Combining the two expressions above gives

$$(S_o/N_o)_{SSB} = (S_i/N_i)_{SSB} = 2(S_i/N_i)_{DSBSC}$$

Hence, with these results we obtain

$$(S_o/N_o)_{SSB} = (S_o/N_o)_{DSBSC} = (S_o/N_o)_{AM}$$

on the basis of the same *average* sideband power.

However, on the basis of *peak* powers transmitted, it can be shown that

$$(S_o/N_o)_{SSB} = 8(S_o/N_o)_{AM}$$

which is an improvement of 9 dB on the AM system.

Example 6.1

Analyse the operation of the envelope and synchronous detectors for use in the demodulation of an amplitude-modulated carrier wave under conditions of very low

University of Strathclyde Libraries
Main Library
CheckOut Receipt

30/04/07
10:59 am

Item:Digital signal analysis / Samuel D.
Stearns, Don R. Hush.
Due date (yyyy-rnm-dd) :

20070611   170000

Thank You for using

Please retain this receipt

signal-to-noise ratio. Derive in each case an expression for the output signal-to-noise ratio.

If the output of both detectors contains mostly noise, explain why the output of the synchronous detector is useful whereas that of the envelope detector is not. (U.L.)

Solution
The incoming AM carrier signal (with random noise) after passing through the band-pass IF filter of the receiver can be represented by

$$v(t) = \{V_c + m(t)\}\sin \omega_c t + n(t)$$

where $m(t)$ is the modulating signal and $n(t)$ is *band-limited* random noise. The latter can be expressed by its in-phase and quadrature phase components $x(t)$ and $y(t)$ respectively, as illustrated in Fig. 6.2(a). Hence

$$v(t) = \{V_c + m(t)\}\sin \omega_c t + x(t)\sin \omega_c t + y(t)\cos \omega_c t$$

or $$v(t) = [\{V_c + m(t)\} + x(t)]\sin \omega_c t + y(t)\cos \omega_c t$$

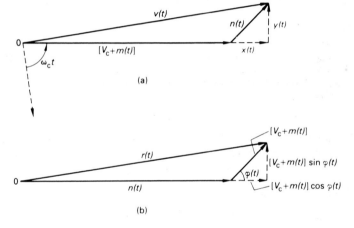

(a)

(b)

Fig. 6.2

Synchronous detector
The signal $v(t)$ is multiplied with a local synchronised carrier $v_c(t) = \sin \omega_c t$ and the detector output $v_d(t)$ is

$$v_d(t) = [\{V_c + m(t)\} + x(t)]\sin^2 \omega_c t + y(t)\sin \omega_c t \cos \omega_c t$$

or $$v_d(t) = [\{V_c + m(t)\} + x(t)]\left[\frac{1 - \cos 2\omega_c t}{2}\right] + y(t)\tfrac{1}{2}(\sin 2\omega_c t)$$

The signal after filtering by a low-pass filter and with the d.c. blocked by an a.c. coupling yields

$$v_o(t) = m(t)/2 + x(t)/2$$

The output $v_o(t)$ contains the modulation explicitly, together with *additive* noise. This result is independent of the input S/N ratio since it does not enter into the analysis. The output S/N ratio is given by

$$(S_o/N_o)_{AM} = \frac{\overline{m^2(t)}}{\overline{x^2(t)}}$$

Envelope detector

For small input S/N ratios, the phasor diagram is shown in Fig. 6.2(b) where the noise component $n(t)$ is taken as the reference phasor. The resultant signal into the detector is given by $r(t)$ where

$$r(t) = n(t) + [V_c + m(t)]\cos\phi(t) + j[V_c + m(t)]\sin\phi(t)$$

and $\phi(t)$ is an arbitrary phase angle which varies continuously from 0 to 2π, due to the random phase of the *band-limited* noise component $n(t)$.

If the output of the detector is $v_o(t)$ we have

$$v_o(t) \simeq n(t) + [V_c + m(t)]\cos\phi(t)$$

since (S_i/N_i) is small. Hence

$$v_o(t) \simeq n(t) + V_c\cos\phi(t) + m(t)\cos\phi(t)$$

The output signal $v_o(t)$ does not contain the modulation *explicitly* as the last term depends on $\cos\phi(t)$ which is varying continuously from 0 to 2π. Hence, the modulation cannot be recovered directly and may be lost when $S_i \ll N_i$. This is called the AM threshold effect and so $(S_o/N_o)_{AM} \simeq 0$.

Comment

The results obtained above show that, in the case of the synchronous detector, there is no threshold effect and the modulation is present even when S_i is small though embedded in noise. In the case of the envelope detector, the modulation is lost due to the threshold effect and only noise is present. Hence, the synchronous detector may still be useful even when $S_i < N_i$.

FM system

The expression for an FM signal entering the discriminator of an FM receiver is

$$v_i(t) = V_c\sin[\omega_c t - (\Delta f/f_m)\cos\omega_m t]$$
$$= V_c\sin[\omega_c t + \phi]$$

where $\phi = -(\Delta f/f_m)\cos\omega_m t$, Δf is the peak frequency deviation, and $\omega_m/2\pi$ is the modulating frequency.

Since a phase shift ϕ corresponds to a frequency deviation of $(1/2\pi)\,d\phi/dt$, the output of the discriminator $v_o(t)$ is proportional to $(1/2\pi)\,d\phi/dt$. Hence

$$v_o(t) = K\frac{1}{2\pi}\frac{d\phi}{dt} = \frac{K}{2\pi}\frac{\Delta f\omega_m}{f_m}\sin\omega_m t$$

or
$$v_o(t) = K\,\Delta f\sin\omega_m t$$

where K is a constant of proportionality. The *average* signal power is S_o where

$$S_o = \frac{K^2 \Delta f^2}{2} \quad \text{watts}$$

To evaluate the effect of random noise, we observe that each noise frequency component f_n will beat with the carrier wave to produce amplitude modulation and angle modulation as illustrated in Fig. 6.3.

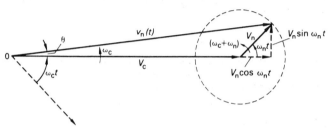

Fig. 6.3

If the noise component has a peak voltage V_n where $V_c \gg V_n$ we obtain

$$v_n(t) = (V_c + V_n \cos \omega_n t) + j V_n \sin \omega_n t$$
$$= V_c [\{1 + (V_n/V_c) \cos \omega_n t\} + j(V_n/V_c) \sin \omega_n t]$$

or $\quad v_n(t) \simeq V_c \sin(\omega_c t + \theta)$

where
$$\theta = \tan^{-1} \frac{(V_n/V_c) \sin \omega_n t}{1 + (V_n/V_c) \cos \omega_n t} \simeq (V_n/V_c) \sin \omega_n t$$

since $V_c \gg V_n$.

The output noise voltage from the discriminator $v_d(t)$ will be proportional to the frequency modulation produced by the noise signal $v_n(t)$ which is related to the phase modulation it produces by

$$v_d(t) = K \frac{1}{2\pi} \frac{d\theta}{dt} = K \frac{V_n}{V_c} f_n \cos \omega_n t$$

The average output noise power δN_o in a bandwidth δf is then given by

$$\delta N_o = \frac{K^2 V_n^2 f_n^2}{2 V_c^2}$$

Previously it was shown that $N_0 \delta f = V_n^2/2$ and $P_c = V_c^2/2$ is the average carrier power. Hence

$$\delta N_o = \frac{K^2 (N_0 \delta f) f_n^2}{2 P_c}$$

The total noise power output N_o in an IF bandwidth $\pm B$ Hz then becomes

$$N_o = \int_{-B}^{+B} \frac{K^2 (N_0 \delta f) f_n^2}{2P_c} = \frac{K^2 N_0}{2P_c} \int_{-B}^{+B} f_n^2 \, df$$

or

$$N_o = \frac{K^2 N_0 B^3}{3P_c}$$

Hence

$$(S_o/N_o)_{FM} = \frac{K^2 \Delta f^2}{2} \left/ \frac{K^2 N_0 B^3}{3P_c} \right. = 3\left(\frac{\Delta f}{B}\right)^2 \frac{P_c}{2N_0 B}$$

or

$$(S_o/N_o)_{FM} = 3m_f^2 \frac{P_c}{2N_0 B}$$

where $m_f = \Delta f/B$ is the modulation index or deviation ratio of the system.

To compare this result with an AM system, assume that $m = 1$ in an AM system corresponds to 100 % deviation in an FM system. The output noise volts in an AM system is the familiar rectangular distribution shown in Fig. 6.4(a) while the FM noise output is triangular. The corresponding noise powers are shown in Fig. 6.4(b).

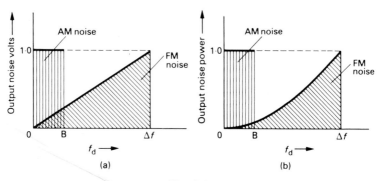

Fig. 6.4

Previously it was shown that the AM output S/N ratio from an envelope detector is given by

$$(S_o/N_o)_{AM} = \frac{m^2 P_c}{2N_0 B} = \frac{P_c}{2N_0 B} \qquad \text{(for } m = 1\text{)}$$

The corresponding FM output S/N ratio was derived as

$$(S_o/N_o)_{FM} = 3m_f^2 \frac{P_c}{2N_0 B}$$

Hence
$$\frac{(S_o/N_o)_{\text{FM}}}{(S_o/N_o)_{\text{AM}}} = \frac{3m_f^2 P_c}{2N_0 B} \bigg/ \frac{P_c}{2N_0 B} = 3m_f^2$$

In particular, if $\Delta f = 75\,\text{kHz}$ and $B = 15\,\text{kHz}$ we have $m_f = 5$ and so the signal-to-noise improvement due to FM is $3 \times 25 = 75$ or 19 dB. The factor of 75 can be increased further by the use of pre-emphasis at the transmitter and de-emphasis at the receiver. It can be shown* that this amounts to about 4 dB, giving an overall S/N improvement compared to AM of 23 dB.

6.2   S/N ratios

Typical graphs of input and output S/N ratios are shown in Fig. 6.5, assuming the same input noise bandwidth for the various systems. At low values of S_i/N_i, the AM systems appear to be the best, but for S_i/N_i above the 10 dB threshold, FM is superior to the AM systems. To achieve large values of S_o/N_o, however, PCM appears to offer the best advantage. As a comparison, the ideal system shown is still about 8 to 10 dB better than the FM or PCM system.

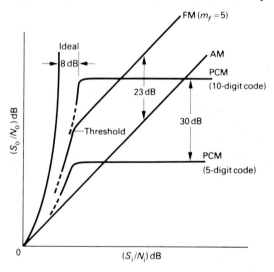

Fig. 6.5

6.3   Pulse systems[35,36]

The most familiar pulse systems used are pulse amplitude modulation (PAM), pulse position modulation (PPM), and pulse code modulation (PCM).

* See F. R. Connor, *Modulation*, Edward Arnold (1982).

PAM system
Pulse amplitude modulation is normally employed in the early stages of PPM or PCM systems since it is easy to multiplex PAM pulses. It is generally not used as the final system, however, since the S/N ratio obtainable is not as good as those of the other pulse systems.

It can be shown that PAM gives results very similar to those obtained previously for AM. Since noise directly affects the amplitude of the pulses, it appears as direct amplitude modulation in the system, as with the modulating signal. Hence, we have

$$(S_o/N_o)_{PAM} = 2(S_i/N_i)_{PAM}$$

PPM system
In pulse position modulation, the modulating voltage causes a time displacement of the pulse. To evaluate the S/N ratio, assume that the maximum time displacement due to the modulation is t_0 and so the peak signal volts is equal to Kt_0 where K is a constant of proportionality. Hence, the average signal power at the output of the detector is

$$S_o = K^2 t_0^2/2$$

The effect of noise in the signal is to alter the time displacement which leads to an error ε, as shown in Fig. 6.6. The rms noise volts causing the error ε produces an rms time displacement Δt such that

$$\frac{\varepsilon}{V_c} = \frac{\Delta t}{t_r} \qquad (\text{if } V_c \gg \varepsilon)$$

where V_c is the peak pulse volts and t_r is the rise-time of the leading pulse edge which carries the modulation.

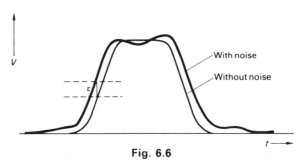

Fig. 6.6

The output rms noise volts is thus $K\,\Delta t$ and the average output noise power at the detector is

$$N_o = K^2 \,\Delta t^2$$

Hence
$$(S_o/N_o)_{PPM} = \frac{K^2 t_0^2}{2} \times \frac{1}{K^2 \Delta t^2} = \frac{t_0^2}{2\Delta t^2}$$

For a reasonably sharp leading edge, the bandwidth B of the filter required to demodulate the pulse train is $B = 1/2t_r$ which yields

$$(S_o/N_o)_{PPM} = 2t_0^2 V_c^2 B^2/\varepsilon^2$$

Now $V_c^2/\varepsilon^2 = C_i/N_i$ where C_i is the *peak* carrier power and N_i is the *mean* noise power, hence

$$(S_o/N_o)_{PPM} = 2t_0^2 B^2 (C_i/N_i)$$

or
$$\frac{(S_o/N_o)_{PPM}}{(C_i/N_i)} = 2t_0^2 B^2$$

and so for a given C_i/N_i ratio the S/N improvement depends on B^2 indicating that a better S/N ratio can be obtained at the expense of an increased bandwidth, i.e. B can be exchanged with S/N, in accordance with the Hartley–Shannon law.

PCM system[37]
In a PCM system, for reasonably good S_i/N_i ratios of around 20 dB or more, the *overriding* noise effect is due to quantisation, i.e. quantisation noise. Its evaluation depends on an estimate of the error introduced by quantisation.

Consider a signal of V volts (peak to peak) quantised into q equal levels (uniform quantisation) with a level spacing of ΔV volts. Any *signal* level, such as $(l+\varepsilon)$, will be associated with an uncertainty or error ε volts such that $-\Delta V/2 < \varepsilon < \Delta V/2$ where l is at the middle of the spacing, as indicated in Fig. 6.7.

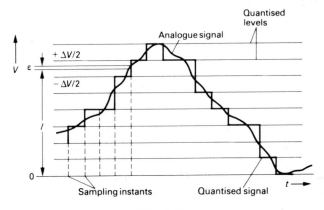

Fig. 6.7

The error ε can take on all possible values between $-\Delta V/2$ and $+\Delta V/2$ and may be considered as due to *added* noise in the signal. Hence, the mean-square value of the error gives the mean-square value of 'quantisation noise'. To calculate it, assume that over a long period of time all levels have an equal probability of occurrence and so the occurrence of any level is the same. Hence, we obtain

$$\overline{\varepsilon^2} = \frac{1}{\Delta V} \int_{-\Delta V/2}^{+\Delta V/2} \varepsilon^2 \, d\varepsilon = \frac{\Delta V^2}{12}$$

or

$$N_o = \frac{\Delta V^2}{12} \quad \text{watts} \qquad \text{(for a 1 } \Omega \text{ load)}$$

To calculate the signal power for q levels spaced ΔV volts apart we have

$$V = (q-1)\Delta V \quad \text{volts}$$

Assuming further that bipolar pulses are used (since less power is consumed), the pulse heights are $\pm \Delta V/2, \pm 3\Delta V/2, \ldots, \pm (q-1)\Delta V/2$. For equal probability of occurrence of all levels in a long message, we obtain the average signal power as

$$S_o = \frac{1}{(q/2)} \left[(\Delta V/2)^2 + 3(\Delta V/2)^2 + \ldots + \{(q-1)\Delta V/2\}^2 \right]$$

or

$$S_o = \frac{\Delta V^2}{2q} \left[1^2 + 3^2 + 5^2 + \ldots + (q-1)^2 \right]$$

This may be written as

$$S_o = \frac{\Delta V^2}{2q} \left[\{1^2 + 2^2 + 3^2 + \ldots + (q-1)^2\} - 2^2 \left\{ 1^2 + 2^2 + \ldots + \left(\frac{q-2}{2}\right)^2 \right\} \right]$$

Now

$$\sum_{n=1}^{n=m} n^2 = \frac{m(m+1)(2m+1)}{6}$$

from which we obtain

$$1^2 + 2^2 + \ldots + (q-1)^2 = \frac{q(q-1)(2q-1)}{6}$$

$$1^2 + 2^2 + \ldots + \left(\frac{q-2}{2}\right)^2 = \frac{q(q-1)(q-2)}{4 \times 6}$$

Hence

$$S_o = \frac{\Delta V^2}{12q} \left[q(q-1)(2q-1) - q(q-1)(q-2) \right]$$

$$= \frac{\Delta V^2}{12}(q-1)[(2q-1) - (q-2)]$$

$$= \frac{\Delta V^2}{12}(q-1)(q+1)$$

$$= \frac{\Delta V^2}{12}(q^2-1)$$

or $\qquad S_o \simeq \frac{\Delta V^2}{12} q^2 \qquad$ (for $q \gg 1$)

Hence, the signal-to-noise ratio due to quantisation noise becomes

$$(S_o/N_o)_{PCM} \simeq \frac{\Delta V^2}{12}(q^2-1) \bigg/ \frac{\Delta V^2}{12}$$

$$\simeq (q^2-1)$$

or $\qquad (S_o/N_o)_{PCM} \simeq q^2 \qquad$ (for $q \gg 1$)

The result depends on the square of the number of levels used and so a large number of levels is required for a large S/N ratio. As an example, if $q = 128$, $(S_o/N_o) \simeq (128)^2 \simeq 42\,\mathrm{dB}$. This requires the use of a 7-digit code since $2^7 = 128$ levels.

6.4   Communication capacity[38]

According to the Hartley–Shannon law of information, the communication capacity C of a system with a bandwidth B and a signal-to-noise ratio S/N is given by

$$C = B \log_2 (1 + S/N) \text{ bits/s}$$

This rate of information transmission may be regarded as the ideal if it is assumed that the error rate is less than 1 in 10^5 bits/s. Comparing this with a binary PCM system, we observe that, for a sampling frequency of $2W$ where W is the highest modulating frequency and q quantised levels are used, the amount of information H transmitted is given by

$$H = \log_2 q \text{ bits}$$

For a sampling frequency of $2W$, at least one pulse is sent in each sampling period and so the total number of pulses sent per second is $n = 2W$. Hence

$$C = H' = nH = 2W \log_2 q \text{ bits/s}$$

or $\qquad C = B \log_2 q^2 \text{ bits/s}$

since $B = 2W$ is the system bandwidth.

Previously it was shown that

$$S_o = \frac{\Delta V^2}{12}(q^2-1)$$

or
$$q^2 = 1 + \frac{12}{\Delta V^2} S_o$$

where S_o is the average output signal power and ΔV is the interval spacing between levels.

Due to noise, there will be an error in distinguishing between any two adjacent levels unless the spacing is large enough. If σ is the rms noise voltage, the minimum spacing ΔV is given by $\Delta V = k\sigma$ where k is a constant of proportionality whose value depends on the error rate assumed. Hence, we obtain the mean noise power N_o in a $1\,\Omega$ load as $N_o = \sigma^2$ with

$$C = B \log_2\left(1 + \frac{12}{\Delta V^2} S_o\right)$$

or
$$C = B \log_2\left(1 + \frac{12}{k^2}\frac{S_o}{N_o}\right)$$

Comparing this result with the ideal system, we observe that a binary PCM system requires $(k^2/12)$ times as much signal power as the ideal.

To evaluate the constant k, it can be shown that, for an error rate of 1 in 10^5, the value of k is about 9. Hence, $k^2/12 \simeq 7$ or about 8 dB which means that the PCM system requires about 7 times as much signal power as the ideal system or is about 8 dB from the ideal system. Its efficiency is thus about 14 % compared with the ideal system. The communication rates of some PCM systems are compared with the ideal in Fig. 6.8.

Example 6.2
Show that, in a binary PCM system, the output signal-to-noise ratio increases exponentially as the transmission bandwidth of the system.

Solution
The output S/N ratio of a binary PCM system was obtained previously as

$$(S_o/N_o)_{PCM} \simeq q^2$$

where q is the number of quantised levels.

For an n-pulse group, we require to transmit $2nB$ pulses per second where B is the highest frequency component in the signal and $2B$ is the sampling frequency. If t is the width of each pulse then $t = 1/2nB$. Also, from telegraphy, it is known that the fundamental frequency $f = 1/2t$. Hence, the *transmission* bandwidth B_t required is given by

$$B_t = 1/2t = nB$$

or
$$n = B_t/B$$

Also, for an n-pulse group we have $2^n = q$ which yields

$$(S_o/N_o)_{PCM} \simeq q^2 \simeq 2^{2n}$$

or
$$(S_o/N_o)_{PCM} \simeq 2^{2(B_t/B)}$$

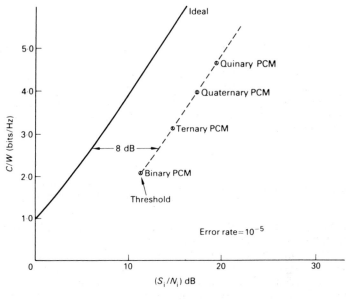

Fig. 6.8

The output S/N ratio therefore increases exponentially since it varies as the *exponent* (B_t/B). Hence, a PCM system can be made superior to other systems by increasing the ratio B_t/B.

6.5 Digital systems[39]

In a digital data system, communication is by means of the two symbols 0 and 1. The symbols are encoded for transmitting messages and a typical example is the teleprinter code used in telegraphy. However, for every symbol transmitted, the receiver must make a choice between the two symbols and so the probability of an error occurring due to noise is a useful criterion for comparing various types of digital systems. It is shown in Appendix L that the probability of error is given by

$$P_e = \frac{1}{2}\text{erfc}\left[\frac{E(1-\rho)}{2N_0}\right]^{\frac{1}{2}}$$

where erfc signifies the complimentary error function, E is the energy per bit transmitted, ρ is the correlation coefficient, and N_0 is the noise power spectral density.

Coherent systems

In amplitude-shift-keying (ASK), the signal is on when symbol 1 is transmitted and is off when symbol 0 is transmitted. Hence, the signal waveforms transmitted are

$$s_1(t) = A \sin \omega t \qquad \text{(for symbol 1)}$$
$$s_0(t) = 0 \qquad \text{(for symbol 0)}$$

There is no correlation between the signals, i.e. $\rho = 0$, and the energy in one of the bits is zero. Hence, we obtain

$$P_e = \tfrac{1}{2}\text{erfc}\,(E/4N_0)^{\frac{1}{2}}$$

if the frequency and phase of the transmitted signal are known at the receiver.

In frequency-shift-keying (FSK), two different frequencies f_0 and f_1 are used for the symbols and we have

$$s_1(t) = A \sin \omega_1 t \qquad \text{(for symbol 1)}$$
$$s_0(t) = A \sin \omega_0 t \qquad \text{(for symbol 0)}$$

and detection is by means of two matched filters. If the frequencies are fairly well separated, the two signals are orthogonal, i.e. $\rho = 0$, and we have

$$P_e = \tfrac{1}{2}\text{erfc}\,(E/2N_0)^{\frac{1}{2}}$$

a result which is less than the previous case if the frequency and phase of each signal are known at the receiver and if the bit ratio E/N_0 remains unchanged.[*] The main disadvantage of ASK over FSK is the need for automatic gain control to overcome fading effects at the receiver.

In binary phase-shift-keying (PSK), two different signal phases are used which are 180° apart such that

$$s_1(t) = A \sin \omega t \qquad \text{(for symbol 1)}$$
$$s_0(t) = -A \sin \omega t \qquad \text{(for symbol 0)}$$

and the two signals are identical but opposite in phase, i.e. $\rho = -1$, and we obtain

$$P_e = \tfrac{1}{2}\text{erfc}\,(E/N_0)^{\frac{1}{2}}$$

which is the minimum value for a given E/N_0 if the frequency and phase are known at the receiver and so it is the *optimum* system.

A variation of PSK, which requires no coherent reference phase at the receiver, is differential phase-shift-keying (DPSK). It uses the previous transmitted bit as the reference for the subsequent bit. Since the previous bit is

---

[*] Coherent ASK and FSK perform equally well if the *average* bit ratio E/N_0 is the same.

contaminated with additive noise, the performance of DPSK is inferior to PSK by about 2 dB. Moreover, bit errors tend to occur in pairs because adjacent bits are related in phase. It can be shown that the error probability is given by

$$P_e = \tfrac{1}{2}e^{-E/N_0}$$

To conserve power or bandwidth, other forms of signalling may be used which employ multi-amplitude, multi-frequency, or multi-phase methods. Of these, quadrature phase-shift-keying (QPSK) is often used with the four phase values of 45°, 135°, 270°, and 315°. Its performance is similar to that of PSK for large values of E/N_0, but it requires only half the bandwidth of the PSK system.

Non-coherent systems
Both the ASK and FSK systems may be used without a known reference signal at the receiver. This is achieved by using envelope detectors and making a decision on which output is the largest. Since the output is affected by both in-phase and out-of-phase noise components, such non-coherent systems lead to some degradation in performance. In the case of non-coherent FSK, the error probability is given by

$$P_e = \tfrac{1}{2}e^{-E/2N_0}$$

which is 3 dB poorer than coherent DPSK.

Bit error rate (BER)
It has been shown above that the probability of bit error (BER) depends essentially on the ratio E/N_0. The bit error rate is of primary importance in digital data systems and, typically, is of the order of 10^{-5} or less.

To reduce errors in a bit stream, the encoded binary data is transmitted with additional *check* digits in some form of code. This increases the bit rate but reduces the ratio E/N_0 for a given transmitter power. Consequently, the error rate would have *increased* were it not for the check digits which also correct the additional errors produced. Hence, there is an overall reduction of bit error rate from 10^{-5} to 10^{-7} in a typical case.

A summary of the various values of BER is given in Table 6.1 and a typical set of graphs showing the relationship between BER and E/N_0 is shown in Fig. 6.9 for the various systems.

6.6   Satellite systems[40, 41]

The performance of a satellite communication system can be described by its uplink and downlink power budgets. Since it is fairly easy to transmit high power to a satellite in a geostationary orbit from a high-gain ground antenna, the main limitation in performance of the satellite system is determined by its *downlink* power budget. This is especially true, because the limited size and mass

Table 6.1

System	Type	BER
ASK	Coherent	$\frac{1}{2}\,\mathrm{erfc}\,(E/4N_0)^{\frac{1}{2}}$
FSK	Coherent	$\frac{1}{2}\,\mathrm{erfc}\,(E/2N_0)^{\frac{1}{2}}$
PSK	Coherent	$\frac{1}{2}\,\mathrm{erfc}\,(E/N_0)^{\frac{1}{2}}$
DPSK	Coherent	$\frac{1}{2}\,e^{-E/N_0}$
QPSK	Coherent	$\simeq \frac{1}{2}\,\mathrm{erfc}\,(E/N_0)^{\frac{1}{2}}$
ASK	Non-coherent	$\simeq \frac{1}{2}\,e^{-E/4N_0}$
FSK	Non-coherent	$\frac{1}{2}\,e^{-E/2N_0}$

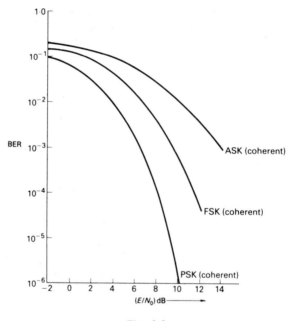

Fig. 6.9

of the satellite severely restrict its transmitter power and antenna gain. Hence, it is important to optimise the downlink power budget for any given satellite to ground communications.

For the purposes of analysis, it will be assumed that the calculations are based on a typical case situation which is likely to occur in practice. For the uplink budget, the critical parameter to be considered is the effective isotropic radiated power (EIRP) at the satellite which is important in achieving maximum power output from its transmitter. It is given by the expression

$$\text{EIRP} = \frac{P_T G_T}{L_F L_S}$$

or
$$\text{EIRP} = P_T + G_T - L_F - L_S \ \text{dB}$$

where P_T = transmitter power
L_F = feeder and diplexer loss
L_S = free space loss
G_T = transmitting antenna gain

Given below are typical values of the relevant parameters at an operating frequency of 6 GHz for a geostationary satellite orbiting at a distance of about 36 000 km above the earth. The losses include the large space loss (201 dB) and the miscellaneous losses due to transmitter ageing (1 dB), antenna pointing error (2 dB), and rain attenuation margin (2 dB).

Transmitter power	23 dBW
Transmitting antenna gain	60 dB
EIRP	83 dBW
Free space loss $(4\pi d^2/\lambda^2)$	−201 dB
Satellite antenna gain	28 dB
Miscellaneous losses	−5 dB
Received carrier power	−95 dBW
Satellite noise power	−126 dBW
Carrier-to-noise power density ratio	97 dBHz
Carrier-to-noise power ratio	31 dB

The critical downlink parameter is the figure of merit G/T since it directly determines the carrier-to-noise power density ratio (C/N_0), which is the ultimate criterion at the receiver's demodulator for analogue signals, or the bit energy-to-noise power density (E/N_0) for digital signals. Hence, it is easily shown that

$$\frac{C}{N_0} = \frac{\text{EIRP} \times G_R}{kT_s L_s L_m M}$$

or
$$\frac{C}{N_0} = \text{EIRP} + G_R - kT_s - L_s - L_m - M \ \text{dBHz}$$

and
$$\frac{C}{N} = \frac{C}{N_0 W} \ \text{dB}$$

where C = carrier power at the demodulator
N_0 = noise power spectral density
N = noise power
EIRP = effective isotropic radiated power
G_R = receiving antenna gain
k = Boltzmann's constant
T_s = system noise temperature
L_s = free space loss
L_m = miscellaneous losses
M = margin for multiple carriers
W = relevant bandwidth

Typical figures for a ground station receiving a signal at 4 GHz from a geostationary satellite are given below. It is convenient here to consider the parameter (C/N_0) for analogue signals since it is independent of the system bandwidth used. For digital signals, the ratio (E/N_0) is easily determined from

$$\frac{C}{N_0} = \frac{C\tau}{N_0} \times \frac{1}{\tau} = \frac{ER}{N_0}$$

or

$$\frac{E}{N_0} = \frac{C}{N_0 R} \,\mathrm{dB}$$

where τ is the bit duration, E is the energy per bit, and R is the bit rate.

Satellite EIRP	18 dBW
Free space loss	−197 dB
Receiving antenna gain	60 dB
System noise temperature	70 K
G/T ratio	41·5 dB/K
Miscellaneous losses	−5 dB
Received carrier power	−124 dBW
Noise power density	−210 dBW/Hz
Carrier-to-noise power density ratio	86 dB Hz
Carrier-to-noise power ratio	20 dB
Bit rate (digital systems)*	$8·5 \times 10^6$ bits/s
Energy-to-noise power density ratio	17 dB
BER	1×10^{-4}

Here, the system noise temperature includes the antenna noise temperature and that of the subsequent receiver chain, the losses include the large space loss (197 dB) and the miscellaneous losses due to atmospheric attenuation (2 dB), antenna aperture efficiency (1 dB), and mispointing and polarisation loss (2 dB).

* 132 channels

The optimum design parameters which determine the performance of the system are essentially trade-offs between various competing factors, such as the quality of communication required, availability of power, choice of antenna gain, and the amount of noise in the system, especially at the receiver.

Example 6.3
(a) Describe with the aid of block diagrams the arrangement of a microwave satellite communication system that incorporates an active artificial earth satellite in geostationary orbit.
(b) Enumerate the system parameters and discuss a typical signal-to-noise budget (balance sheet of trade-offs).   (C.E.I.)

Solution
(a) A typical system employing a satellite in geostationary orbit is the *Intelsat* system which provides point-to-point international communication in the frequency band around 6 GHz (uplink) and 4 GHz (downlink) via a ground station. The ground station employs a large steerable dish of about 30 m diameter with a minimum G/T ratio of 40·7 dB/K and recent satellites use global horn beams or spot paraboloid beams, orthogonally polarised. A typical schematic diagram is shown in Fig. 6.10 and the main application is for telephone and television traffic, using FM/FDMA techniques. More recently, TDM/TDMA has been used and future development is towards PCM digital techniques.

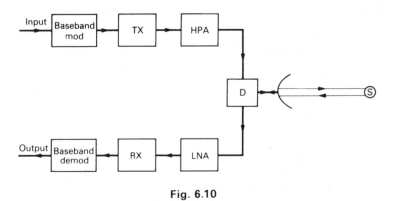

Fig. 6.10

The ground station equipment consists of a transmitter chain and receiver chain, which are isolated by a diplexer. In the uplink path, FM modulators drive the TX and the IF of 70 MHz is up-converted to 6 GHz at low level. The high-power amplifier feeding the antenna uses klystrons or TWTs. At the satellite, the signal is amplified and down-converted to 4 GHz before transmission back to earth.

In the downlink, the received signal with a bandwidth up to 36 MHz lies within a 500 MHz band at 4 GHz and so low-noise wideband amplifiers are essential. Typically, a cooled paramp at 15 K is followed by a receiver which down-converts the signal to 70 MHz. The FM demodulators are conventional, though some may employ FMFB techniques, while the TX and RX use conventional baseband FM equipment.

(b)   The performance of a satellite communication system is determined by various design parameters which are trade-offs between the quality of information desired, the availability of power, gain, and bandwidth, the various noise contributions in the system, and the considerable losses due to the great distances involved. The system is characterised by uplink and downlink power budgets which were given earlier.

The critical uplink parameter is the EIRP of the ground station. Various losses occur between the ground station and satellite, which include the large space loss at 6 GHz (201 dB) and miscellaneous losses of about 6 dB. The received carrier power at the satellite determines its S/N or C/N_0 ratios and should be sufficiently high to overcome the degradation in the downlink.

The ultimate criterion which determines the quality of information is the S/N ratio at the demodulator in the ground station receiver. For an FM system, this should be better than 10 dB and so the critical downlink parameter is the G/T ratio of the ground station. For a given satellite capability, the important parameters which determine the downlink power budget are the EIRP of the *satellite*, the large space loss at 4 GHz (197 dB), and miscellaneous losses of about 9 dB. A typical S/N ratio at the demodulator is about 20 dB for the relevant bandwidth considered.

Problems

1   Three coins are tossed in succession. What is the probability that the tosses
     result in
     (a)   two tails,
     (b)   at least two tails?
2   A bag contains two white beads and three black beads. Another bag
     contains six white beads and four black beads. Two beads only are drawn
     in succession, one from each bag. What is the probability that the beads
     drawn are
     (a)   both white,
     (b)   both black,
     (c)   one white and one black?
3   A Gaussian probability density function is given by

$$p(x) = \frac{e^{-(x-\alpha)^2/2\sigma^2}}{\sigma\sqrt{2\pi}}$$

     where σ is the standard deviation and α is a constant. Determine the first
     moment (average value) of the distribution and calculate the percentage
     area under the curve which lies within $\pm\sigma$ of the average value.
4   A probability density function is defined by $p(x) = Ke^{-2|x|}$ where K is a
     constant. For the normalised function, determine (a) the value of K and
     (b) the variance, when $-\infty < x < +\infty$.
5   A random signal is defined by $v(t) = V_0$ when $-\delta/2 < t < +\delta/2$ and
     $v(t) = 0$ at all other instants. Determine its autocorrelation function.
6   If the autocorrelation function of a random signal is given by
     $R(\tau) = \tau[(\sin \omega_0\tau/2)/(\omega_0\tau/2)]^2$, determine its power spectral density.
7   Discuss, very briefly, the origins of 'noise' in a communication system.
         A microphone having an internal impedance of $100\,\Omega$ (resistive) is
     matched to the internal impedance of an amplifier by means of a step-up
     transformer. The microphone is capable of producing an open-circuit
     signal voltage of $20\,\mu$V. If the bandwidth of the system is $15\,$kHz and the
     noise figure of the amplifier is $7\,$dB, calculate the signal-to-noise ratio at
     the output terminals. State any assumptions you make. What would be the
     effect of removing the transformer from the input circuit and connecting
     the microphone directly to the amplifier input?   (U.L.)

8   Derive an expression for the overall noise factor of a combination of three two-port networks, connected in cascade, expressed in terms of the individual noise factors and gains of the individual networks

A receiving system consists of a preamplifier connected through a length of cable to a main receiver. The noise factor of the preamplifier is 6 dB, while the corresponding values for the cable section and receiver are 8 and 13 dB. Given the attenuation of the cable is 8 dB, calculate the minimum gain required in the preamplifier if the overall noise factor of the system is not to exceed 9 dB.   (U.L.)

9   (a)   Describe briefly what is meant by the noise factor of a radio receiver.
    (b)   Two amplifiers, connected in series, are matched in impedance and bandwidth. The first has a gain of 16 dB and a noise factor of 6 dB, the second has a gain of 10 dB and a noise factor of 3 dB. Calculate (i) the noise factor of the combination and (ii) the noise factor if the order of the amplifiers is reversed.   (C.G.L.I.)

10 A parabolic antenna has a noise temperature of 60 K and it is connected by a length of waveguide to a parametric preamplifier. The waveguide has a loss of 1 dB while the preamplifier has an effective noise temperature of 77 K and a gain of 20 dB. If a receiver is used with the preamplifier and its noise factor is 10 dB, what is the effective noise temperature of the system?

11 A communication receiver system has an input stage with a noise temperature of 100 K and a loss of 5 dB. This is followed by three IF amplifier modules, each with a gain of 10 dB, a bandwidth of 6 MHz, and a noise figure of 1·3 dB. With the system matched throughout, calculate
    (a)   the system noise figure,
    (b)   the equivalent noise temperature of the system when connected to an aerial with a noise temperature of 50 K,
    (c)   the smallest usable input signal power if the communication recognition system operates correctly for an output signal-to-noise ratio > 1,
    (d)   the effect on (a), (b), and (c) of adding a preamplifier with a noise temperature of 100 K and gain of 20 dB. The receiver system is at a temperature of 290 K.   (C.E.I.)

12 A signal $s(t)$ is a triangular pulse of the form

$$s(t) = Kt \qquad 0 \leqslant t \leqslant T$$

$$s(t) = 0 \qquad \text{at all other values of time}$$

where K is a constant. Determine the output of a filter matched to this signal.

If Gaussian white noise of zero mean value and noise spectral density N_0 (positive frequencies only) is added to the signal, what is the maximum signal-to-noise ratio at the output of the matched filter?

13 In a pulse radar system, the observed signal is received in the presence of Gaussian noise of zero mean and unit variance. Assuming the received signal is of 2 volt amplitude, determine for a Neyman–Pearson receiver

the probability of detection when the false alarm probability is set at 0·2.

14 A 30 channel PCM system with uniform quantisation and a 7-bit binary code has an output bit rate of 1·5 Mbits/s. Determine
   (a) the maximum information bandwidth over which satisfactory operation is possible,
   (b) the output signal-to-quantising noise ratio for an input sinusoidal signal at a frequency of 3 kHz and maximum design amplitude.

15 Calculate the signal-to-noise ratio for a sinusoidal signal quantised into M levels given that the total mean-square quantising noise voltage $\sigma^2 = 0.083(\Delta v)^2$ where Δv is the step size. What assumption has been made about the quantisation? Hence, estimate the number of digits per character required in a PCM system carrying the above signal if the quality has to be satisfactory for the telephone system.

   Discuss whether speech processed in a similar manner has the same quality. Explain how the signal-to-noise ratio for speech can be improved by analogue or digital signal processing.   (C.E.I.)

16 A coherent binary data system uses on–off pulses varying in amplitude from 0 to V volts. The probability of a 0 or 1 in the presence of Gaussian noise is the same. For a peak signal power to average noise power ratio of 13 dB, calculate the probability of error. Also, for a probability of error of 10^{-5}, determine the signal-to-noise amplitude threshold required.

17 An FSK communication channel transmits binary information at a bit rate of 100 kbits/s in the presence of Gaussian noise with a spectral density of 10^{-19} W/Hz. If the signal is transmitted with a peak voltage level of 1 volt, determine for a probability of error of 10^{-4} the path loss of the channel for bit by bit detection with
   (a) incoherent FSK,
   (b) coherent FSK.

18 Explain why communications satellites fulfil an important role in worldwide communications. Mention, particularly, aspects of the role that cannot be readily fulfilled by alternative systems.

   A satellite in a geostationary orbit at 35800 km has a 4 GHz downlink transmitter which feeds 25 watts into an antenna with 20 dB gain. The ground station receiving system has a total noise figure of 1·5 dB. Calculate the antenna gain necessary at the ground station to maintain a 30 dB input signal-to-noise ratio over a 12 MHz signal band.   (C.E.I.)

19 The parameters of a satellite-to-ship link are as follows

Satellite RF transmitter power per channel	$+ P_s$ dB W
Satellite aerial gain relative to isotropic	$+ 17$ dB
Free space path loss at 1·5 GHz	$- 189$ dB
Propagation margin at 5° elevation for 99 % of the time	$+ 5$ dB
Ship aerial gain relative to isotropic	$+ G_r$ dB
RF input power to ship receiver	$- 152$ dB W

  (a) From the above information, sketch a graph of satellite transmitter power per channel P_s, against appropriate values of ship aerial gain G_r, to give the required input level to the ship receiver.
  (b) The maritime satellite transmitter can supply only a total of 50 W RF power to the satellite aerial. Indicate on the graph the minimum value of ship aerial gain that would permit the satellite to transmit 10 separate channels simultaneously.
  (c) State *two* factors which govern the upper limit of ship aerial gain.
                                                                    (C.G.L.I.)
20 Draw up a balance sheet for a 12–14 GHz satellite link using a 1 metre parabolic dish at the satellite and a 3 metre dish at the ground station terminal. Evaluate the signal-to-noise ratio obtained
  (a) at the input to the satellite receiver,
  (b) at the output of the ground station receiver, assuming voice transmission over a 12 MHz bandwidth. The uplink and downlink transmitter power for a geostationary orbit is 300 W in each direction.

Answers

1   $\frac{3}{8}$, $\frac{1}{2}$

2   $\frac{6}{25}$, $\frac{6}{25}$, $\frac{13}{25}$

3   α,   68 %

4   $K = 1$, $\sigma^2 = \frac{1}{2}$

5   $R(\tau) = V_0^2 \delta [1 - |\tau|/\delta]$ $(|\tau| < \delta)$
   $R(\tau) = 0$ $(|\tau| \geqslant \delta)$

6   $S(f) = 1 - |f|/f_0$ $(|f| < f_0)$
   $S(f) = 0$ $(|f| \geqslant f_0)$

7   42 dB,   35 dB

8   15 dB

9   6·05 dB,   3·62 dB

10   265 K

11   (a) 7·5 dB,   (b) 1404 K,   (c) 0·66 μW,
   (d) 1·44 dB,   163·5 K,   0·019 μW

12   $K^2 T^3/3$, $2K^2 T^3/3N_0$

13   0·627

14   3·57 kHz,   43·8 dB

15   $S/N_{\cdot q} = 10 \log(1\cdot 5\, M^2)$
Linear quantisation is assumed.
7 digits
The speech quality is improved by using non-linear quantisation (analogue processing) or by using more quantised levels (digital processing).

16   $1\cdot 27 \times 10^{-2}$,   threshold ratio = 4·25

17   124·6 dB,   125·6 dB

18   54·6 dB

20   35 dB $\pm$ 3 dB margin,   27 dB $\pm$ 3 dB margin

References

1   STEINBERG, J. L. and LEQUEX, J. *Radio Astronomy*. McGraw-Hill (1963).
2   KRASSNER, G. N. and MICHAELS, J. Y. *Introduction to Space Communication Systems*. McGraw-Hill (1964).
3   LATHI, B. P. *Introduction to Random Signals and Communication Theory*. International Textbook Co. (1968).
4   WHALEN, A. D. *Detection of Signals in Noise*. Academic Press (1971).
5   LINDGREN, B. W. and McELRATH, G. W. *Introduction to Probability and Statistics*. Macmillan (1969).
6   HOEL, P. G. *Introduction to Mathematical Statistics*. John Wiley (1971).
7   COOPER, G. R. and McGILLEM, C. D. *Probabilistic Methods of Signal and System Analysis*. Holt, Rinehart and Winston (1971).
8   CONNOR, F. R. *Signals*. Edward Arnold (1982).
9   ZIEMER, R. E. and TRANTER, W. H. *Principles of Communications, Systems, Modulation and Noise*. Houghton Mifflin Co. (1976).
10 PANTER, P. E. *Communication Systems Design*. McGraw-Hill (1972).
11 LANGE, F. H. *Correlation Techniques*. Iliffe (1967).
12 JOHNSON, D. A. H. Cross-correlation techniques and their application to communication channel evaluation, *The Radio and Electronic Engineer*, **37**, 315, May 1969.
13 WIENER, N. Generalised harmonic analysis, *Acta Mathematica*, **55**, 117, 1930.
14 KHINTCHINE, A. Korrelationstheorie der statistischen Prozesse, *Math. Annalen*, **109**, 604, 1934.
15 BEAUCHAMP, K. and YUEN, C. *Digital Methods for Signal Analysis*. George Allen and Unwin (1974).
16 SRINATH, M. D. and RAJASEKARAN, P. K. *Introduction to Statistical Signal Processing with Applications*. John Wiley (1979).
17 JOHNSON, J. B. Thermal agitation of electricity in conductors, *Physical Review*, **32**, 97, 1928.
18 NYQUIST, H. Thermal agitation of electric charge in conductors, *Physical Review*, **32**, 110, 1928.
19 DAVENPORT, W. B. and ROOT, W. L. *An Introduction to the Theory of Random Signals and Noise*. McGraw-Hill (1958).
20 NIELSEN, E. G. Behaviour of noise figure in junction transistors, *Proceedings Institute of Radio Engineers*, **45**, 957, 1957.
21 MOTCHENBACHER, C. D. and FITCHEN, F. C. *Low-Noise Electronic Design*. John Wiley (1973).
22 LIECHTI, C. A. Microwave field-effect transistors, *Institute of Electrical and Electronic Engineers Transactions*, **MTT-24**, 279–300, June 1976.
23 VAN DER ZIEL, A. and CHENETTE, E. R. Noise in solid state devices, *Advances in Electronics and Electron Physics*, **46**, 313, Academic Press (1978).

24 FRIIS, H. T. Noise figures of radio receivers, *Proceedings Institute of Radio Engineers*, **45**, 957, 1957.
25 GUPTA, M. S. (ed.) *Electrical Noise, Fundamentals and Sources*. IEEE Press (1977).
26 CUCCIA, C. L. Ultra low-noise parametric amplifiers in communication satellite earth terminals, *Advan es in Microwaves*, **7**, 175, 1971.
27 DAGLISH, H. N. *et al. Low-Noise Microwave Amplifiers*. The University Press, Cambridge (1968).
28 KENNEDY, G. *Electronic Communication Systems*. McGraw-Hill (1977).
29 MANLEY, J. M. and ROWE, H. E. *Proceedings Institute of Radio Engineers*, **44**, 904, July 1956.
30 POSNER, R. D. Design and specification considerations of GaAs FET low-noise amplifiers, *Microwave Journal*, 69, May 1979.
31 KREUTEL, R. W. and PACHOLDER, A. O. Measurement of gain and noise temperature of a satellite communications ground station, *Microwave Journal*, 12, October 1969.
32 LONGLEY, S. R. Design and application information for broadband solid-state noise sources. *Mullard Technical Communications No. 138*, April 1978.
33 CARLSON, A. B. *Communication Systems*. McGraw-Hill (1975).
34 TAUB, H. and SCHILLING, D. L. *Principles of Communication Systems*. McGraw-Hill (1971).
35 STREMLER, F. G. *Introduction to Communication Systems*. Addison-Wesley (1977).
36 PEEBLES, P. Z. *Communication System Principles*. Addison-Wesley (1976).
37 OLIVER, B. M. and PIERCE, J. R. The philosophy of PCM, *Proceedings Institute of Radio Engineers*, **36**, 1324, November 1948.
38 SHANNON, C. E. Communication in the presence of noise, *Proceedings Institute of Radio Engineers*, **37**, 10, 1949.
39 LAWTON, J. G. Comparison of binary data transmission systems, *Proceedings National Convention on Military Electronics*, 54–61, 1958.
40 MARTIN, J. *Communication Satellite Systems*. Prentice-Hall (1978).
41 TURNER, L. W. (ed.) *Electronic Engineers' Reference Book*. Newnes-Butterworth (1976).

Appendices

Appendix A: Set theory

Some basic ideas of set theory will first be presented before illustrating their application to probability theory.

Definitions

Set and subsets
Elements or objects which have a common property are called a *set*. Typical examples are the numbers 1 to 6 on a die or the letters of an alphabet. The total set of elements is called a universal set S and a smaller set of elements is called a *subset*. A set or subset may be represented by an area on a Venn diagram or Karnaugh map as illustrated in Fig. A.1.

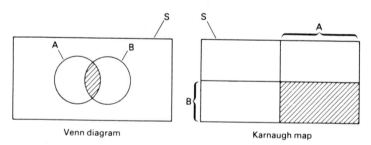

Venn diagram   Karnaugh map

Fig. A.1

Union
The union of two subsets A and B is written as $A \cup B$. It is another subset of S which contains the area of A or B or both and is shown in Fig. A.2(a).

Intersection
The intersection of two subsets A and B is written as $A \cap B$. It is the subset which contains the area common to both A and B as shown in Fig. A.2(b).

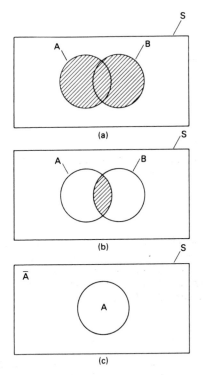

Fig. A.2

Complement
The complement of subset A is written as $\bar{A}$. It is the subset of S which does not contain the area of subset A and is shown in Fig. A.2(c).

Application
The basis of probability theory can be established using set theory. This will be illustrated by considering some probabilities.

$P(A+B)$
If A and B are discrete probabilities (not mutually exclusive) then $P(A)$ can be represented by subset A, $P(B)$ by subset B, and $P(AB)$ by the subset common to both A and B. Hence, $P(A+B)$ is given by $A \cup B$ since the shaded area in Fig. A.2(a) is subset A plus subset B *minus* the area common to A and B. Hence, we obtain

$$P(A+B) = P(A) + P(B) - P(AB)$$

which is in accord with probability theory.

$P(AB)$

The joint probability of A and B (not mutually exclusive) is given by the subset common to both subset A and subset B. This is the shaded area shown in Fig. A.2(b) and so it is given by $A \cap B$.

$P(A|B)$

Since $P(A|B)$ denotes the conditional probability of A given B has already occurred, we need to know when subset A and subset B occur together. Hence, $P(A|B)$ is related to the shaded area $P(AB)$ in Fig. A.2(b). To ensure that this area is always less than 1, it must be divided by subset B to *normalise* it. Hence, we obtain

$$P(A|B) = \frac{P(AB)}{P(B)}$$

which was given by Bayes' theorem.

Appendix B: Error function

The error function was defined in Section 2.6 as

$$\operatorname{erf} x = \frac{2}{\sqrt{\pi}} \int_0^x e^{-u^2} \, du$$

and values of x from 0 to 3·00 are given in Table A.1. To obtain the complementary error function, use the relationship

$$\operatorname{erfc} x = 1 - \operatorname{erf} x$$

Appendix C: Power and voltage spectral densities

Power spectral density

To obtain the value of P_{av}, consider first the periodic signal $x(t)$ shown in Fig. A.3(a). Assume that a 'sample' of this signal is extracted over the period $-T/2 < t < +T/2$ to produce the pulse-like signal $X(t)$ shown in Fig. A.3(b). If $F(\omega)$ is the Fourier transform of $X(t)$ then

$$X(t) = \frac{1}{2\pi} \int_{-\infty}^{+\infty} F(\omega) e^{j\omega t} \, d\omega$$

Also, we have

$$\frac{1}{T} \int_{-T/2}^{+T/2} X^2(t) \, dt = \frac{1}{T} \int_{-T/2}^{+T/2} X(t) \left\{ \frac{1}{2\pi} \int_{-\infty}^{+\infty} F(\omega) e^{j\omega t} \, d\omega \right\} dt$$

where the second quantity $X(t)$ is replaced by the previous expression for $X(t)$

Table A.1

x	erf x	x	erf x
0	0	1·50	0·9661
0·05	0·0563	1·55	0·9716
0·10	0·1124	1·60	0·9763
0·15	0·1679	1·65	0·9803
0·20	0·2227	1·70	0·9837
0·25	0·2763	1·75	0·9866
0·30	0·3286	1·80	0·9890
0·35	0·3793	1·85	0·9911
0·40	0·4283	1·90	0·9927
0·45	0·4754	1·95	0·9941
0·50	0·5205	2·00	0·9953
0·55	0·5663	2·05	0·9962
0·60	0·6038	2·10	0·9970
0·65	0·6420	2·15	0·9976
0·70	0·6778	2·20	0·9981
0·75	0·7111	2·25	0·9985
0·80	0·7421	2·30	0·9988
0·85	0·7706	2·35	0·9991
0·90	0·7969	2·40	0·9993
0·95	0·8208	2·45	0·9994
1·00	0·8427	2·50	0·9995
1·05	0·8624	2·55	0·9996
1·10	0·8802	2·60	0·9997
1·15	0·8961	2·65	0·9998
1·20	0·9103	2·70	0·99986
1·25	0·9229	2·75	0·99989
1·30	0·9340	2·80	0·99992
1·35	0·9437	2·85	0·99994
1·40	0·9522	2·90	0·99995
1·45	0·9596	2·95	0·99996
		3·00	0·99997

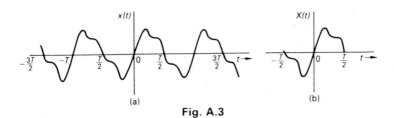

(a) (b)

Fig. A.3

in terms of $F(\omega)$. Hence, rearranging the order of integration then yields

$$\frac{1}{T}\int_{-T/2}^{+T/2} X^2(t)\,\mathrm{d}t = \frac{1}{2\pi T}\int_{-\infty}^{+\infty} F(\omega)\,\mathrm{d}\omega\left[\int_{-T/2}^{+T/2} X(t)\mathrm{e}^{\mathrm{j}\omega t}\mathrm{d}t\right]$$

or
$$\frac{1}{T}\int_{-T/2}^{+T/2} X^2(t)\,\mathrm{d}t = \frac{1}{2\pi T}\int_{-\infty}^{+\infty} F(\omega)\,\mathrm{d}\omega \int_{-\infty}^{+\infty} X(t)\mathrm{e}^{\mathrm{j}\omega t}\,\mathrm{d}t$$

since $X(t)$ is zero over the intervals $-\infty < t < -T/2$ and $+T/2 < t < +\infty$. Also, we have

$$\int_{-\infty}^{+\infty} X(t)\mathrm{e}^{\mathrm{j}\omega t}\,\mathrm{d}t = F^*(\omega)$$

where $F^*(\omega)$ is the conjugate of $F(\omega)$ such that $F(\omega)\,F^*(\omega) = |F(\omega)|^2$. Hence

$$\frac{1}{T}\int_{-T/2}^{+T/2} X^2(t)\,\mathrm{d}t = \frac{1}{2\pi}\int_{-\infty}^{+\infty} \frac{|F(\omega)|^2}{T}\,\mathrm{d}\omega$$

As more 'samples' of $x(t)$ are removed over time intervals of T and added to $X(t)$, the signal $X(t)$ will eventually resemble the periodic signal $x(t)$ and so in the limit when $T \to \infty$ we obtain

$$\lim_{T \to \infty}\frac{1}{T}\int_{-T/2}^{+T/2} X^2(t)\,\mathrm{d}t = \lim_{T \to \infty}\frac{1}{T}\int_{-T/2}^{+T/2} x^2(t)\,\mathrm{d}t = \lim_{T \to \infty}\frac{1}{2\pi}\int_{-\infty}^{+\infty} \frac{|F(\omega)|^2}{T}\,\mathrm{d}\omega$$

The quantity in the centre is the average power of the periodic signal $x(t)$ and so we obtain

$$P_{\mathrm{av}} = \lim_{T \to \infty}\frac{1}{2\pi}\int_{-\infty}^{+\infty} \frac{|F(\omega)|^2}{T}\,\mathrm{d}\omega = \frac{1}{2\pi}\int_{-\infty}^{+\infty} S(\omega)\,\mathrm{d}\omega$$

from the *definition* of $S(\omega)$. Hence, by inspection, we obtain

$$S(\omega) = \lim_{T \to \infty}\frac{|F(\omega)|^2}{T}$$

Voltage spectral density

The voltage spectral density $S_{\mathrm{v}}(f)$ is defined by

$$\int_{-\infty}^{+\infty} S_{\mathrm{v}}(f)\,\mathrm{d}f = \overline{v_{\mathrm{n}}^2}$$

where $\overline{v_{\mathrm{n}}^2}$ is the mean-square noise voltage. In the case of thermal noise, the noise spectrum is constant over a finite bandwidth B. Hence

$$\int_{-B}^{+B} S_{\mathrm{v}}(f)\,\mathrm{d}f = 4kTBR$$

or
$$S_{\mathrm{v}}(f)\int_{-B}^{+B}\mathrm{d}f = 4kTBR$$

Hence
$$S_v(f)\,2B = 4kTBR$$
or
$$S_v(f) = 2kTR$$

Appendix D: Wiener–Khintchine theorem

For a 'sample' function $X(t)$ where $X(t)$ is a single pulse signal, we have from Appendix C

$$S(\omega) = \lim_{T \to \infty} \frac{|F(\omega)|^2}{T}$$

where $F(\omega)$ is the Fourier transform of $X(t)$.

Now, if $X(t)$ is real, we have from transform properties that

$$F(\omega) = \int_{-\infty}^{+\infty} X(t)\mathrm{e}^{-j\omega t}\,\mathrm{d}t$$

and
$$F(\omega) = F(-\omega)$$

Also, $|F(\omega)|^2 = F(\omega)F(-\omega)$

or
$$|F(\omega)|^2 = \int_{-\infty}^{+\infty} X(t_1)\mathrm{e}^{-j\omega t_1}\,\mathrm{d}t_1 \int_{-\infty}^{+\infty} X(t_2)\mathrm{e}^{j\omega t_2}\,\mathrm{d}t_2$$

where $t_2 = t_1 - \tau$. Hence

$$S(\omega) = \lim_{T \to \infty} \frac{\displaystyle\int_{-T/2}^{+T/2} x(t_1)\mathrm{e}^{-j\omega t_1}\,\mathrm{d}t_1 \int_{-T/2}^{+T/2} x(t_2)\mathrm{e}^{j\omega t_2}\,\mathrm{d}t_2}{T}$$

where $x(t)$ replaces $X(t)$ since only a 'sample' of $x(t)$ in the interval $-T/2 < t < +T/2$ is being considered.

Interchanging the order of integration and substituting $t_2 = t_1 - \tau$ yields

$$S(\omega) = \lim_{T \to \infty} \frac{\displaystyle\int_{-T/2}^{+T/2} \int_{-T/2}^{+T/2} x(t_1)x(t_1-\tau)\,\mathrm{d}t_1\,\mathrm{e}^{-j\omega\tau}\,\mathrm{d}\tau}{T}$$

Now
$$\lim_{T \to \infty} \frac{1}{T}\int_{-T/2}^{+T/2} x(t_1)x(t_1-\tau)\,\mathrm{d}t_1 = R(\tau)$$

Hence
$$S(\omega) = \int_{-\infty}^{+\infty} R(\tau)\mathrm{e}^{-j\omega\tau}\,\mathrm{d}\tau$$

and by taking the inverse transform we obtain

$$R(\tau) = \frac{1}{2\pi}\int_{-\infty}^{+\infty} S(\omega)\mathrm{e}^{j\omega\tau}\,\mathrm{d}\omega$$

Appendix E: Narrowband noise

If Gaussian white noise is passed through a narrow bandpass filter with a bandwidth B which is very much less than the centre frequency f_c of the filter, the output from the filter is known as narrowband noise. It has the spectrum shown in Fig. A.4(a) which can be approximated by a finite number of noise components spaced Δf apart where $\Delta f \to 0$ and this is shown in Fig. A.4(b).

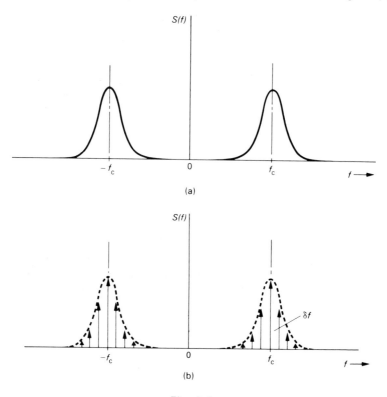

(a)

(b)

Fig. A.4

Each pair of delta functions, such as $\delta(f+f_c)$ and $\delta(f-f_c)$, can be represented by a sine-wave function of arbitrary phase and the sum of all such spectral components yields the time function $n(t)$ of narrowband noise. Hence, we obtain

$$n(t) = \lim_{\Delta f \to 0} \sum_{0}^{n} a_n \{ \sin[(\omega_c + 2\pi n \Delta f)t + \theta_n] \}$$

where a_n is the amplitude of the nth frequency component and θ_n is its arbitrary phase. Expanding this expression then yields

$$n(t) = \lim_{\Delta f \to 0} \sum_0^n a_n \{\sin \omega_c t \cos (2\pi n \Delta f t + \theta_n) + \cos \omega_c t \sin (2\pi n \Delta f t + \theta_n)\}$$

or
$$n(t) = x(t) \sin \omega_c t + y(t) \cos \omega_c t$$

where
$$x(t) = \lim_{\Delta f \to 0} \sum_0^n a_n \cos (2\pi n \Delta f t + \theta_n)$$

and
$$y(t) = \lim_{\Delta f \to 0} \sum_0^n a_n \sin (2\pi n \Delta f t + \theta_n)$$

The components $x(t)$ and $y(t)$ are called the in-phase and quadrature phase components if a sine function is used as a reference for the centre frequency component f_c. Both $x(t)$ and $y(t)$ are Gaussian distributions with the same mean and variance as $n(t)$.

The expression for $n(t)$ can be written in polar form by the substitution

$$x(t) = R(t) \cos \phi(t)$$
$$y(t) = R(t) \sin \phi(t)$$

where
$$R(t) = \sqrt{x^2(t) + y^2(t)}$$

and
$$\phi(t) = \tan^{-1}[y(t)/x(t)]$$

Hence
$$n(t) = R(t) \sin [\omega_c t + \phi(t)]$$

where the amplitude $R(t)$ varies with a Rayleigh distribution and the phase of $\phi(t)$ varies uniformly over the interval 0 to 2π. Hence, $n(t)$ resembles a sine wave which is amplitude-modulated and phase-modulated randomly as shown in Fig. A.5.

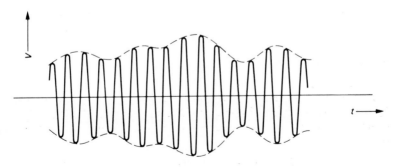

Fig. A.5

Appendix F: Matched filter

The output of a linear filter is given by

$$s_o(t) = s_i(t) + s_n(t)$$

where $s_i(t)$ is the input signal and $s_n(t)$ is assumed to be white noise. The output signal-to-noise ratio to be maximised at time $t = t_0$ is given by

$$\frac{S}{N} = \frac{[s_i(t_0)]^2}{s_n^2(t)}$$

If the transfer function of the filter is $H(\omega)$ and the noise power spectral density function is $N_0/2$ (assuming positive and negative frequencies) we obtain

$$\frac{S}{N} = \frac{\left(\dfrac{1}{4\pi}\right)^2 \left| \displaystyle\int_{-\infty}^{+\infty} H(\omega) S(\omega) e^{j\omega t_0} \, d\omega \right|^2}{\left(\dfrac{N_0}{4\pi}\right) \displaystyle\int_{-\infty}^{+\infty} |H(\omega)|^2 \, d\omega}$$

or

$$\frac{S}{N} = \frac{\left| \displaystyle\int_{-\infty}^{+\infty} |H(\omega)S(\omega) e^{j\omega t_0} \, d\omega \right|^2}{N_0 \pi \displaystyle\int_{-\infty}^{+\infty} |H(\omega)|^2 \, d\omega}$$

where $S(\omega)$ is the Fourier transform of the input signal $s_i(t)$.

From Schwarz's inequality we have

$$\left| \int_{-\infty}^{+\infty} F_1(\omega) F_2(\omega) \, d\omega \right|^2 \leqslant \int_{-\infty}^{+\infty} |F_1(\omega)|^2 \, d\omega \int_{-\infty}^{+\infty} |F_2(\omega)|^2 \, d\omega$$

and if $F_1(\omega) = H(\omega)$ and $F_2(\omega) = S(\omega) e^{j\omega t_0}$ we obtain

$$\left| \int_{-\infty}^{+\infty} H(\omega) S(\omega) e^{j\omega t_0} \, d\omega \right|^2 \leqslant \int_{-\infty}^{+\infty} |H(\omega)|^2 \, d\omega \int_{-\infty}^{+\infty} |S(\omega)|^2 \, d\omega$$

and the equality sign holds when

$$H(\omega) = k S^*(\omega) e^{-j\omega t_0}$$

where k is an arbitrary constant and $S^*(\omega)$ is the complex conjugate of $S(\omega)$. Substituting for $H(\omega)$, with $k = 1$ for convenience, then yields

$$\frac{S}{N} = \frac{\displaystyle\int_{-\infty}^{+\infty} |S(\omega)|^2 \, d\omega}{N_0 \pi}$$

Furthermore, if E is the signal energy we have

$$E = \int_{-\infty}^{+\infty} S^2(t)\,dt = \frac{1}{2\pi} \int_{-\infty}^{+\infty} |S(\omega)|^2\,d\omega$$

and substituting this expression into $(S/N)_{max}$ yields

$$\left(\frac{S}{N}\right)_{max} = \frac{2E}{N_0}$$

a result which depends on the signal energy but is independent of its waveform. If the impulse response of the matched filter is $h(t)$, the Fourier transform of an impulse $\delta(t)$ is unity and we have

$$H(\omega) = \frac{F[h(t)]}{F[\delta(t)]} = F[h(t)]$$

where $H(\omega) = S^*(\omega)e^{-j\omega t_0}$. Taking the inverse transform then yields

$$h(t) = \frac{1}{2\pi} \int_{-\infty}^{+\infty} S^*(\omega)e^{-j\omega t_0} \times e^{j\omega t}\,d\omega$$

or

$$h(t) = \frac{1}{2\pi} \int_{-\infty}^{+\infty} S^*(\omega)e^{j\omega(t-t_0)}\,d\omega$$

and the r.h.s. is simply the inverse transform expression for the signal $s^*(t - t_0)$. Hence, it follows that

$$h(t) = s^*(t - t_0) = s(t_0 - t)$$

and the impulse response of the matched filter is the input signal delayed and time-reversed. For $h(t)$ to be real, or for the filter to be physically realisable, we must have $t \leqslant t_0$, i.e. all the signal energy in $s(t)$ must be received before the decision is made at $t = t_0$.

Correlation detector

If the input signal to the matched filter is $s(t)$, the output signal $s_0(t)$ is given by the convolution integral as

$$s_0(t) = \int_{-\infty}^{+\infty} s(\tau)h(t - \tau)\,d\tau$$

where τ is an arbitrary variable, $h(t) = s(t_0 - t)$ is the impulse response of the matched filter, and $h(t - \tau) = s(t_0 - t + \tau)$. Hence

$$s_0(t) = \int_{-\infty}^{+\infty} s(\tau)s(t_0 - t + \tau)\,d\tau$$

and at time $t_0 = 0$ we obtain

$$s_o(t) = \int_{-\infty}^{+\infty} s(\tau) s(\tau - t) \, d\tau$$

which is the autocorrelation function of $s(t)$. Hence, the output of the matched filter is the same as that obtained by cross-correlating the input signal $s(t)$ (plus noise) with the expected signal $s(t)$ and sampling the output at time $t = t_0$, as was illustrated in Fig. 3.12(a).

Appendix G: Decision theory

Let the transmitted signal be $x(t)$ and the received signal be $y(t)$ which contains additive Gaussian noise $n(t)$. Hence

$$y(t) = x(t) + n(t)$$

where $x(t)$ and $y(t)$ may have either discrete or continuous values depending on the type of signal transmitted.

Bayes' criterion

Assume, for simplicity, the discrete binary channel shown in Fig. A.6, where $x_1(t)$ or $x_2(t)$ are the transmitted signals and $y_1(t)$ or $y_2(t)$ are the received signals respectively.

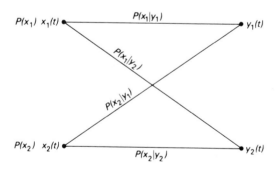

Fig. A.6

The various probabilities involved are the *a priori* probabilities $P(x_1)$ and $P(x_2)$ and the conditional probabilities $P(x_1 \mid y_1)$, $P(x_1 \mid y_2)$, $P(x_2 \mid y_2)$, and $P(x_2 \mid y_1)$. Furthermore, in the decision making process, let H_1 or H_2 be the two hypotheses corresponding to the transmission of $x_1(t)$ or $x_2(t)$. The various costs involved will depend on the choice of hypothesis and the costs are denoted as follows.

C_{11}—if H_1 is chosen when H_1 is true
C_{12}—if H_1 is chosen when H_2 is true
C_{21}—if H_2 is chosen when H_1 is true
C_{22}—if H_2 is chosen when H_2 is true

For *any* received signal y_k, a decision is made according to hypothesis H_1 or H_2, which will involve the average conditional costs $C(H_1 \mid y_k)$ or $C(H_2 \mid y_k)$ since the cost depends on H_1 or H_2 and y_k. For example, if $x_2(t)$ is transmitted we obtain

$$C(H_1 \mid y_k) = C_{11} P(x_1 \mid y_k) + C_{12} P(x_2 \mid y_k)$$

$$C(H_2 \mid y_k) = C_{21} P(x_1 \mid y_k) + C_{22} P(x_2 \mid y_k)$$

and for minimum costs we must have

$$C(H_2 \mid y_k) < C(H_1 \mid y_k) \quad \text{or} \quad C(H_1 \mid y_k) > C(H_2 \mid y_k)$$

Hence $\quad C_{11} P(x_1 \mid y_k) + C_{12} P(x_2 \mid y_k) > C_{21} P(x_1 \mid y_k) + C_{22} P(x_2 \mid y_k)$

or $$\frac{P(x_2 \mid y_k)}{P(x_1 \mid y_k)} > \frac{C_{21} - C_{11}}{C_{12} - C_{22}}$$

From Bayes' theorem, the conditional probabilities $P(x_1 \mid y_k)$ and $P(x_2 \mid y_k)$ are given by

$$P(x_1 \mid y_k) = \frac{P(y_k \mid x_1) P(x_1)}{P(y_k)}$$

$$P(x_2 \mid y_k) = \frac{P(y_k \mid x_2) P(x_2)}{P(y_k)}$$

with $$\frac{P(y_k \mid x_2)}{P(y_k \mid x_1)} = \frac{P(x_1) P(x_2 \mid y_k)}{P(x_2) P(x_1 \mid y_k)}$$

and using the inequality above yields

$$\frac{P(y_k \mid x_2)}{P(y_k \mid x_1)} > \frac{P(x_1)(C_{21} - C_{11})}{P(x_2)(C_{12} - C_{22})}$$

The l.h.s. is denoted by the likelihood ratio $L(y_k)$ and the r.h.s. is called the threshold value L_t. Since hypothesis H_2 should be chosen when $x_2(t)$ is transmitted, i.e. when $L(y_k) > L_t$, hypothesis H_1 should be chosen when $L(y_k) < L_t$. Hence, symbolically, for any received signal y_k which is of the form $y(t)$ we have

$$L[y(t)] \underset{H_1}{\overset{H_2}{\gtrless}} L_t$$

Neyman–Pearson criterion

This criterion is used when the *a posteriori* probabilities only are known. For the analysis, assume that $x(t)$ is discrete, e.g. 0 or 1, but that $y(t)$ is continuous with additive Gaussian noise $n(t)$. Let the transmitted signal be $x_2(t)$ instead of the alternative $x_1(t)$. The conditional probabilities for any received signal $y(t)$ are given by probability distribution functions as follows.

$$P(H_1 \mid x_1) = \int_{-\infty}^{y_t} p_1(y \mid x_1) \, \mathrm{d}y$$

$$P(H_2 \mid x_1) = \int_{y_t}^{+\infty} p_1(y \mid x_1) \, \mathrm{d}y$$

$$P(H_1 \mid x_2) = \int_{-\infty}^{y_t} p_2(y \mid x_2) \, \mathrm{d}y$$

$$P(H_2 \mid x_2) = \int_{y_t}^{+\infty} p_2(y \mid x_2) \, \mathrm{d}y$$

where y_t is the assumed threshold value which may be a particular level of voltage, as shown in Fig. A.7.

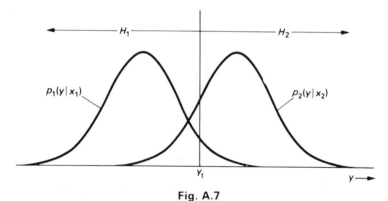

Fig. A.7

It is convenient here to define a detection probability P_D and a false alarm probability P_F such that

$$P_D = P(H_2 \mid x_2) = \int_{y_t}^{+\infty} p_2(y \mid x_2) \, \mathrm{d}y$$

which is the probability of detecting $x_2(t)$ (the assumed signal transmitted) when it is present at the receiver, and so choosing H_2 correctly. Similarly, we have

$$P_F = P(H_2 | x_1) = \int_{y_t}^{+\infty} p_1(y|x_1)\,\mathrm{d}y$$

which is the probability that $x_2(t)$ was *not* transmitted, yet deciding it is present at the receiver, and so choosing H_2 incorrectly.

To apply this criterion, it is usual to decide on a maximum acceptable false alarm probability P_F and then to maximise the detection probability P_D. The resulting values of P_D and P_F when plotted yield the *receiver operating characteristics* shown in Fig. A.8. Parameter d is the signal-to-noise *voltage* ratio at the output of the receiver.

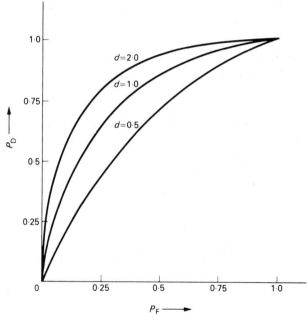

Fig. A.8

For any given false alarm probability P_F, a corresponding threshold value y_t is determined to yield the maximum value for the detection probability P_D. If $y(t) > y_t$, hypothesis H_2 is chosen and, if $y(t) < y_t$, hypothesis H_1 is chosen.

To show the relationship with Bayes' criterion, we obtain the derivatives of

P_D and P_F with respect to y_t which yield

$$\frac{dP_D/dy_t}{dP_F/dy_t} = \frac{dP_D}{dP_F} = \frac{p_2(y_t|x_2)}{p_1(y_t|x_1)}$$

from the expressions for P_D and P_F which were given earlier. By analogy, we define a threshold value L_t as

$$L_t = \frac{p_2(y_t|x_2)}{p_1(y_t|x_1)}$$

or

$$L_t = \frac{dP_D}{dP_F}$$

and so the threshold value L_t for the likelihood ratio is given by the value of the slope of the receiver operating characteristic at any point. However, the threshold value will be different for each criterion.

Appendix H: Estimation theory

In many decision and estimation problems, observations are made of a continuous waveform rather than of discrete values to which Gaussian white noise is an added component. To solve such problems, assume that a set of m observations is made of the waveform and that, in the limit, $m \to \infty$. This is similar to the sampling technique and will be applied to the case of a binary communication system in which the signal waveforms are either $s_0(t)$ or $s_1(t)$ and are non-zero over the interval 0 to T.

A typical observation $z(t)$ is the sum of the signal sample function $s(t)$ and a Gaussian white noise process $n(t)$. Hence, $z(t) = s(t) + n(t)$ which can be written conveniently as $z_i = s_{ki} + n_i$ for the ith observation where $k = 0$ or 1 and $i = 1, 2, \ldots, m$.

Assuming that the noise is band-limited with a noise spectral density $S(\omega) = N_0/2$ over the band $-W < f < W$, the autocorrelation function $R(\tau)$ is given by

$$R(\tau) = N_0 W \frac{\sin x}{x}$$

where $x = 2\pi W\tau$. The zeros of $R(\tau)$ occur when $x = j\pi$ where $j = \pm 1, \pm 2$, etc., i.e. $\tau = 1/2\,W, 1/4\,W$, etc., and the signal samples are uncorrelated if the samples occur at time intervals of $T_s = 1/2\,W$ and so $m = 2WT$ samples.

Defining the likelihood function of the samples as the conditional probability function $p(z|s_k)$ we have

$$p(z|s_k) = \prod_{i=1}^{i=m} p(z_i|s_{ki}) \qquad (k = 0 \text{ or } 1)$$

and if the variance of the noise is denoted by σ^2 we obtain by the Wiener–Khintchine relationship

$$\sigma^2 = R(0) = N_0 W$$

or

$$\sigma^2 = N_0/2T_s$$

Hence

$$p(z \mid s_k) = \prod_{i=1}^{i=m} \frac{1}{\sqrt{2\pi}\,\sigma} \exp\left[\frac{-(z_i - s_{ki})^2}{2\sigma^2} \right]$$

or

$$p(z \mid s_k) = \left(\frac{1}{2\pi\sigma^2} \right)^{m/2} \exp\left[\sum_{i=1}^{i=m} -\frac{1}{2\sigma^2} (z_i - s_{ki})^2 \right]$$

and substituting for σ^2 yields

$$p(z \mid s_k) = \left(\frac{T_s}{\pi N_0} \right)^{m/2} \exp\left[\sum_{i=1}^{i=m} -\frac{T_s}{N_0} (z_i - s_{ki})^2 \right]$$

In the limit, as $m \to \infty$, $T_s \to 0$, but $mT_s = T$ is finite, and the summation becomes an integration over the time interval T. Hence, we obtain

$$p(z \mid s_k) = A \exp\left[-\frac{1}{N_0} \int_0^T \{z(t) - s(t)\}^2 \, dt \right]$$

where A is a constant of no particular significance.

Cramér–Rao bound

An estimator is said to be *unbiased* if the expected value $\hat{\theta}$ of any parameter is equal to its true value θ. The estimation procedure involves the conditional probability function $p(z \mid \theta)$ and for an unbiased estimate we have

$$E[\hat{\theta} \mid \theta] = \theta$$

and

$$E[\hat{\theta} - \theta] = 0$$

Hence

$$\int_{-\infty}^{+\infty} (\hat{\theta} - \theta) p(z \mid \theta) \, dz = 0$$

which after differentiation with respect to θ and a rearrangement of the integration yields

$$\int_{-\infty}^{+\infty} \left[(\hat{\theta} - \theta) \frac{\partial p(z \mid \theta)}{\partial \theta} \, dz - p(z \mid \theta) \, dz \right] = 0$$

or

$$\int_{-\infty}^{+\infty} (\hat{\theta} - \theta) \frac{\partial p(z \mid \theta)}{\partial \theta} \, dz = \int_{-\infty}^{+\infty} p(z \mid \theta) \, dz$$

Since
$$\int_{-\infty}^{+\infty} p(z|\theta)\,dz = 1$$

and
$$\frac{\partial p(z|\theta)}{\partial \theta} = p(z|\theta)\frac{\partial \ln p(z|\theta)}{\partial \theta}$$

we obtain
$$\int_{-\infty}^{+\infty} (\hat{\theta}-\theta)p(z|\theta)\frac{\partial \ln p(z|\theta)}{\partial \theta}\,dz = 1$$

or
$$\left\{\int_{-\infty}^{+\infty} (\hat{\theta}-\theta)p(z|\theta)\frac{\partial \ln p(z|\theta)}{\partial \theta}\,dz\right\}^2 = 1$$

From Schwarz's inequality we have
$$\int_{-\infty}^{+\infty} f_1^2(x)\,dx \int_{-\infty}^{+\infty} f_2^2(x)\,dx \geqslant \left[\int_{-\infty}^{+\infty} f_1(x)f_2(x)\,dx\right]^2$$

Substituting for
$$f_1(x) = (\hat{\theta}-\theta)\sqrt{p(z|\theta)}$$

and
$$f_2(x) = \frac{\partial \ln p(z|\theta)}{\partial \theta}\sqrt{p(z|\theta)}$$

yields
$$\int_{-\infty}^{+\infty} (\hat{\theta}-\theta)^2 p(z|\theta)\,dz \int_{-\infty}^{+\infty} \left\{\frac{\partial \ln p(z|\theta)}{\partial \theta}\right\}^2 p(z|\theta)\,dz \geqslant 1$$

where the first integral is the variance $\sigma_{\hat{\theta}}^2 = E[(\hat{\theta}-\theta)^2]$ and the second integral is the expected value of $[\partial \ln p(z|\theta)/\partial \theta]^2$.

Hence
$$\sigma_{\hat{\theta}}^2 \geqslant \frac{1}{E\left[\left\{\dfrac{\partial \ln p(z|\theta)}{\partial \theta}\right\}^2\right]}$$

which is known as the Cramér–Rao bound.

Comment
An unbiased estimator which attains this lower bound is known as a *minimum variance* unbiased estimator or as an *efficient* estimator.

Appendix I: Thermal noise

Nyquist's derivation of the noise power associated with a resistor is based on the consideration of a lossless transmission line terminated at either end in its characteristic impedance $Z_0 = R$, as shown in Fig. A.9(a). In thermal equilibrium at temperature T, the average noise power N from each resistor travels

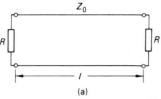
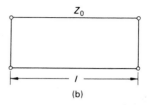

Fig. A.9

along the line as an electromagnetic wave and is completely absorbed at the other end.

If the line is suddenly short-circuited at either end, as in Fig. A.9(b), the noise power from *each* resistor travelling along the line is reflected from each end to produce standing waves. The energy 'trapped' is stored in oscillating modes, as the system becomes a resonator or one-dimensional harmonic oscillator. For such a resonator, the mth mode wavelength λ is given by

$$m\lambda/2 = l$$

or
$$m = 2l/\lambda = 2lf/v$$

where l is the length of the line, f is the mode frequency, and v is the velocity of propagation of the wave.

If the number of oscillating modes Δm occupy a bandwidth $\Delta f = B$, by using increments we obtain

$$\Delta m = 2(l/v)\,\Delta f = 2(l/v)B$$

or
$$2l/v = \Delta m/B$$

The average noise power N from *each* resistor travels for a time $t = l/v$ before reflection and is stored as energy ΔW where

$$\Delta W = 2Nt = 2Nl/v = N\,\Delta m/B$$

From classical theory, it is known that the one-dimensional harmonic oscillator in thermal equilibrium at temperature T is associated with oscillating modes and the energy in each mode is kT where k is Boltzmann's constant. Hence, for Δm modes we have

$$kT\,\Delta m = N\,\Delta m/B$$

or
$$N = kTB \text{ watts}$$

Comment

From quantum theory, the average energy associated with each oscillating mode is modified by Planck's law and we obtain

$$\Delta W = \frac{hf}{e^{hf/kT} - 1}$$

where f is the frequency of oscillation and h is Planck's constant. For frequencies up to about 10^{13} Hz, $\Delta W \simeq kT$ which is the classical value.

Appendix J: Shot noise

The shot noise rms current I_s in a diode is due to the random emission of electrons from the cathode. Each electron arriving at the anode carries a discrete electronic charge e which gives rise to a current pulse $i(t)$ in the anode during the transit time τ, as shown in Fig. A.10(a). The actual shape of the current pulse is immaterial if the time-average interval chosen is such that $\tau \ll T$. Each pulse can be regarded as a Dirac delta function $\delta(t)$ and approximated by a short rectangular pulse, as illustrated in Fig. A.10(b). Hence, we have

$$\int_{-\infty}^{+\infty} \delta(t)\,dt = e$$

i.e. the area of the rectangular pulse is such that $e/\tau \times \tau = e$.

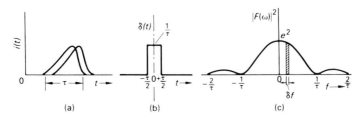

Fig. A.10

If $F(\omega)$ is the Fourier transform of $\delta(t)$ then

$$F(\omega) = \int_{-\infty}^{+\infty} \delta(t)e^{-j\omega t}\,dt = e\frac{\sin \omega\tau/2}{\omega\tau/2}$$

and

$$|F(\omega)|^2 = e^2 \left[\frac{\sin \omega\tau/2}{\omega\tau/2}\right]^2$$

where $|F(\omega)|^2$ is the energy spectral density and is shown in Fig. A.10(c).

From Fig. A.10(c) we observe that if the transit time τ is very small (about 10^{-9} s) then $1/\tau \simeq 10^9$ Hz and the spectral density over a bandwidth $\Delta f = B$ is fairly constant, especially at lower frequencies. Hence, the total energy W in a bandwidth B is given by

$$W = \int_{-\infty}^{+\infty} |F(\omega)|^2\,df = 2\int_0^{+\infty} |F(\omega)|^2\,df$$

or

$$W = 2e^2 B$$

If n electrons arrive at the anode in time T where T is sufficiently large, the average shot noise power in a $1\,\Omega$ load becomes

$$I_s^2 = nW/T = n2e^2B/T$$

and substituting for the average anode current $I_a = ne/T$ yields

$$I_s^2 = 2eI_aB$$

or

$$I_s = \sqrt{2eI_aB}$$

Appendix K: Noise factors

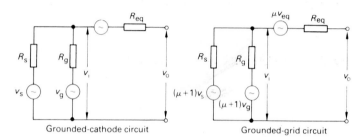

Grounded-cathode circuit   Grounded-grid circuit

Fig. A.11

Grounded-cathode amplifier

The equivalent circuit is shown in Fig. A.11 where R_s is the source resistance, R_g is the grid-leak resistance, and R_{eq} is the equivalent shot noise resistance of the valve. If their rms noise voltages are v_s, v_g, and v_{eq} respectively, the total noise voltage of the amplifier is v_o and for an ideal amplifier it is v_i where v_i is due to the source resistance only. Hence

$$F = \frac{\text{mean-square noise voltage of amplifier}}{\text{mean-square noise voltage due to source}} = \frac{v_o^2}{v_i^2}$$

$$= \frac{4kTB[R_{eq} + R_sR_g/(R_s + R_g)]}{[v_sR_g/(R_s + R_g)]^2}$$

Since $v_s^2 = 4kTBR_s$, we obtain

$$F = \frac{[R_{eq}(R_s + R_g) + R_sR_g](R_s + R_g)}{R_sR_g^2}$$

or

$$F = 1 + R_s/R_g + \frac{R_{eq}}{R_s}[1 + R_s/R_g]^2$$

Grounded-grid amplifier
The equivalent circuit is shown in Fig. A.11. Since the effective amplification factor of the grounded-grid amplifier is $(\mu + 1)$, the rms noise voltages v_s and v_g appear as $(\mu + 1)v_s$ and $(\mu + 1)v_g$ in the equivalent circuit. The grid noise voltage v_{eq}, however, is common to both grid and anode circuits and so its value is $(\mu + 1)v_g - v_g = \mu v_g$ in the equivalent circuit. Hence, we obtain

$$F = \frac{\text{mean-square noise voltage of amplifier}}{\text{mean-square noise voltage due to source}} = \frac{v_o^2}{v_i^2}$$

$$= \frac{\mu^2 4kTBR_{eq} + (\mu + 1)^2 4kTBR_s R_g/(R_s + R_g)}{(\mu + 1)^2 4kTBR_s [R_g/(R_s + R_g)]^2}$$

$$= \frac{[\mu/(\mu + 1)]^2 R_{eq} + R_s R_g/(R_s + R_g)}{R_s [R_g/(R_s + R_g)]^2}$$

or

$$F = 1 + R_s/R_g + \left(\frac{\mu}{\mu + 1}\right)^2 \frac{R_{eq}}{R_s}[1 + R_s/R_g]^2$$

Common-base transistor

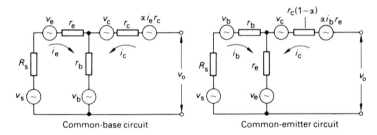

Common-base circuit   Common-emitter circuit

Fig. A.12

The equivalent T-circuit is shown in Fig. A.12 where R_s is the source resistance and r_e, r_b, and r_c are the emitter, base, and collector resistances respectively. The rms noise voltages v_s and v_b are due to thermal noise in the source and base resistances respectively, while v_e and v_c are due to shot noise and partition noise in the emitter and collector regions respectively. Hence, we have for a bandwidth B and absolute temperature T

$$v_s^2 = 4kTBR_s$$
$$v_b^2 = 4kTBr_b$$
$$v_e^2 = I_s^2 r_e^2 = 2kTBr_e$$

since $I_s^2 = 2eI_e B$ and $r_e = kT/eI_e$ if I_e is the d.c. emitter current. Also, the partition noise voltage v_c is given by

$$v_c^2 = 2eI_c\left[1 - \frac{|\alpha|^2}{\alpha_0}\right]Br_c^2 \qquad (|\alpha| = \alpha_0/\{1 + (f/f_{\alpha b})^2\}^{1/2})$$

where e is the electronic charge, I_c is the d.c. collector current, α_0 is the d.c. current-gain factor, α is the current-gain factor at the frequency f, and r_c is the collector resistance. Hence, we obtain

$$F = \frac{\text{mean-square noise voltage of transistor}}{\text{mean-square noise voltage due to source}} = \frac{v_o^2}{v_i^2}$$

where v_i is the rms noise voltage at the output of the transistor due to the source only. Hence

$$v_o^2 \simeq (|\alpha| i_e r_c)^2 + v_c^2 \qquad (\text{if } r_b \ll r_c)$$

and

$$i_e^2 = \frac{v_s^2 + v_e^2 + v_b^2}{(r_e + r_b + R_s)^2}$$

or

$$v_o^2 \simeq \left(\frac{|\alpha| r_c}{r_e + r_b + R_s}\right)^2 (v_s^2 + v_e^2 + v_b^2) + v_c^2$$

Also, with the source noise only we obtain

$$v_i^2 \simeq (|\alpha| r_c)^2\left[\frac{v_s}{r_e + r_b + R_s}\right]^2 \qquad (\text{if } r_b \ll r_c)$$

or

$$v_i^2 \simeq \left(\frac{|\alpha| r_c}{r_e + r_b + R_s}\right)^2 v_s^2$$

Hence

$$F = 1 + \frac{v_e^2}{v_s^2} + \frac{v_b^2}{v_s^2} + \left(\frac{r_e + r_b + R_s}{|\alpha| r_c}\right)^2 \frac{v_c^2}{v_s^2}$$

Now

$$\frac{v_e^2}{v_s^2} = \frac{2kTBr_e}{4kTBR_s} = \frac{r_e}{2R_s} \quad \text{and} \quad \frac{v_b^2}{v_s^2} = \frac{4kTBr_b}{4kTBR_s} = \frac{r_b}{R_s}$$

with

$$\frac{v_c^2}{|\alpha|^2 r_c^2 v_s^2} = \frac{2eI_c[1 - \alpha_0/\{1 + (f/f_{\alpha b})^2\}]Br_c^2}{\alpha_0^2/\{1 + (f/f_{\alpha b})^2\}r_c^2 4kTBR_s}$$

$$= \frac{[1 + (f/f_{\alpha b})^2 - \alpha_0]}{2\alpha_0(kT/eI_e)R_s} \qquad (I_c = \alpha_0 I_e)$$

$$= \frac{(1 - \alpha_0)\left[1 + \left(\dfrac{f/f_{\alpha b}}{\sqrt{1 - \alpha_0}}\right)^2\right]}{2\alpha_0 r_e R_s} \qquad (r_e = kT/eI_e)$$

Substituting the various quantities into the expression for F yields

$$F = 1 + \frac{r_e}{2R_s} + \frac{r_b}{R_s} + \frac{(1-\alpha_0)(R_s + r_e + r_b)^2 \left[1 + \left(\frac{f/f_{\alpha b}}{\sqrt{1-\alpha_0}}\right)^2\right]}{2\alpha_0 r_e R_s}$$

which is illustrated in Fig. A.13.

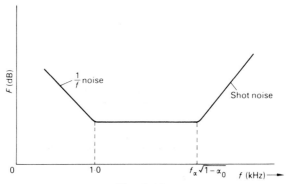

Fig. A.13

Common-emitter transistor
The equivalent T-circuit is shown in Fig. A.12 and, by a similar analysis, it can be shown that the noise factor is the same as that for the common-base circuit if $r_e \ll (1-\alpha)r_c$.

FET noise
The equivalent circuit for the general two-port device with uncorrelated noise sources is shown in Fig. A.14. Here, the source is represented by a current

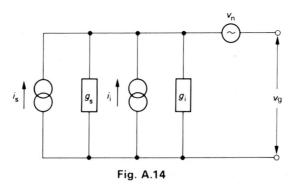

Fig. A.14

generator in parallel with the source conductance g_s, i_i is the noise current generator in parallel with conductance g_i, and v_n is the voltage noise source.

The noise factor is evaluated by considering the ratio of the mean-square voltage $\overline{v_g^2}$, due to all the noise sources, and the mean-square voltage v_s^2, due to the source only. Hence, we have

$$F = \overline{v_g^2}/v_s^2$$

with

$$v_n^2 = 4kTR_nB$$

$$\overline{i_i^2} = 4kTg_iB$$

$$i_s^2 = 4kTg_sB$$

$$v_s^2 = i_s^2/(g_s + g_i)^2$$

where R_n is an equivalent noise resistance, g_i is an equivalent noise conductance at the absolute temperature T, and B is the bandwidth considered. Hence, we obtain

$$\overline{v_g^2} = \frac{4kTg_sB}{(g_s + g_i)^2} + \frac{4kTg_iB}{(g_s + g_i)^2} + 4kTR_nB$$

with

$$F = 1 + \frac{g_i}{g_s} + \frac{R_n(g_i + g_s)^2}{g_s}$$

There is correlation between the channel noise and gate noise and the equivalent circuit for the common-source configuration with neutralised gate-drain capacitance is represented by Fig. A.15. Here, $g_i = g_g + g_c$ since g_g accounts for the uncorrelated gate noise and g_c accounts for the partial correlation.

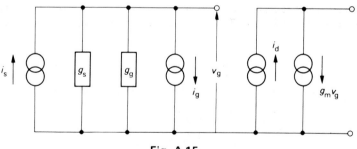

Fig. A.15

For the JFET, $R_n \simeq 0.65/g_m$ where g_m is the low-frequency transconductance, $g_i \simeq g_g$, and g_c is very small compared to g_g, i.e. $g_c \simeq 0$. Hence, by differentiating with respect to g_s and equating to zero, we obtain the condition for F_{min} as

$$g_s = (g_i/R_n + g_i^2)^{1/2} \simeq (g_g/R_n + g_g^2)^{1/2}$$

with
$$F_{min} \simeq 1 + 2R_n g_g + 2\sqrt{R_n g_g + R_n^2 g_g^2}$$

and, typically, when $R_n g_g = 1$, it yields

$$F_{min} \simeq 1 + 2 + 2\sqrt{2} \simeq 6 \quad \text{or} \quad 4\,\text{dB}$$

Similar considerations also apply to the MOSFET, but both R_n and g_g are frequency-dependent. Moreover, there is no shot noise in the gate region and so $i_i = 0$. The noise figure becomes

$$F = 1 + R_n g_{in}$$

and
$$F_{min} \simeq 1 + 0 \cdot 52 \omega C / g_m$$

where g_{in} is the input conductance, C is the gate-to-channel capacitance, and g_m is the forward transconductance.

Appendix L: Probability of error

Let the two possible waveforms received be $s_0(t)$ when 'zero' is transmitted and $s_1(t)$ when 'one' is transmitted such that

$$2E = \int_0^T [s_0^2(t) + s_1^2(t)]\,dt$$

where E is the average signal energy received per symbol of duration T.

We now define a normalised cross-correlation coefficient ρ between the two waveforms such that

$$\rho = \frac{1}{E} \int_0^T s_0(t)s_1(t)dt$$

and then perform a cross-correlation operation between the received signal plus noise waveform $y(t) = s(t) + n(t)$ and both the expected waveforms, coherently detect each, and examine the difference. The decision is then made as to whether the variable ψ is positive or negative where ψ is given by

$$\psi = \int_0^T y(t)[s_0(t) - s_1(t)]dt$$

Since the received signal is a linear combination of the transmitted signal and additive Gaussian noise of *one-sided* noise power density N_0, ψ will also be Gaussian.

If $s_0(t)$ was transmitted then the mean value of ψ is

$$\bar{\psi} = \int_0^T s_0(t)\,[s_0(t) - s_1(t)]\,\mathrm{d}t$$

or

$$\bar{\psi} = E(1 - \rho)$$

and the variance of ψ is given by

$$\sigma^2 = E\left[\left\{\int_0^T n(t)[s_0(t) - s_1(t)]\,\mathrm{d}t\right\}^2\right]$$

with

$$\sigma^2 = \frac{N_0}{2}\int_0^T [s_0(t) - s_1(t)]^2\,\mathrm{d}t$$

or

$$\sigma^2 = EN_0(1 - \rho)$$

The probability density of this Gaussian distribution is given by

$$p(\psi) = \frac{1}{\sigma\sqrt{2\pi}}\exp\left[\frac{-(\psi - \bar{\psi})^2}{2\sigma^2}\right]$$

and the probability of error, i.e. the probability that ψ will be negative when it should be positive, is

$$P_e = \frac{1}{\sigma\sqrt{2\pi}}\int_{-\infty}^0 \exp\left[\frac{-(\psi - \bar{\psi})^2}{2\sigma^2}\right]\mathrm{d}\psi$$

with

$$P_e = \frac{1}{2}\left[1 - \mathrm{erf}\left(\frac{\bar{\psi}}{\sigma\sqrt{2}}\right)\right]$$

and

$$P_e = \frac{1}{2}\left[1 - \mathrm{erf}\left\{\frac{E(1 - \rho)}{2N_0}\right\}^{1/2}\right]$$

or

$$P_e = \frac{1}{2}\,\mathrm{erfc}\left\{\frac{E(1 - \rho)}{2N_0}\right\}^{1/2}$$

Index

Amplifier
    grounded-cathode, 60, 127
    grounded-grid, 60, 128
    FET, 73
    low-noise, 70, 74
amplitude-shift-keying, 94
AM system, 80, 87
attenuator, 63, 67
average value, 10, 15
autocorrelation function, 19, 22, 28, 31, 35
available noise power, 4, 45, 55

Band-limited
    random noise, 83
    white noise, 17, 31
Bayes'
    criterion, 40, 118
    theorem, 8, 110
binary PCM system, 91, 92
binomial distribution, 12
bipolar transistor noise, 52
bit error rate, 95

Cascaded networks, 62
circuit noise, 3, 44
coloured noise, 31, 32
communication capacity, 91
conditional probability, 7
correlation
    coefficient, 20
    detection, 35
    detector, 117
    techniques, 20
Cramér–Rao bound, 41, 123
cross-correlation function, 26

Decision theory, 4, 40, 118

detector
    correlation, 117
    envelope, 80, 84
    synchronous, 83
differential phase-shift-keying, 94
digital systems, 93
Dirac delta function, 30, 126
distribution
    binomial, 12
    Gaussian, 13
    Poisson, 13
    Rayleigh, 15
DSBSC system, 81

Effective noise temperature, 65
ensemble averaging, 17
envelope detector, 80, 84
equivalent noise resistance, 49
ergodicity, 19
error
    function, 15
    PCM system, 89
    PPM system, 88
    rate, 91, 92
estimation theory, 41, 122
excess noise ratio, 78

FET
    amplifier, 73
    noise, 54, 130
FM system, 84, 87
Fourier transform, 27, 110, 113, 126
frequency-shift-keying, 94

Gaussian
    distribution, 13, 35
    white noise, 30